The Legal Technology Guidebook

Kimberly Williams · John M. Facciola
Peter McCann · Vincent M. Catanzaro

The Legal Technology
Guidebook

 Springer

Kimberly Williams
RedShift Legal, Inc.
Brooklyn, NY
USA

and

New York Law School
New York, NY
USA

John M. Facciola
U.S. Federal Judge (ret.)
Washington, DC
USA

and

Georgetown Law School
Washington, DC
USA

Peter McCann
Loop
Philadelphia, PA
USA

and

Drexel University
Philadelphia, PA
USA

Vincent M. Catanzaro
Shook, Hardy & Bacon, L.L.P.
Philadelphia, PA
USA

ISBN 978-3-319-85414-4   ISBN 978-3-319-54523-3   (eBook)
DOI 10.1007/978-3-319-54523-3

Printed on acid-free paper

This Springer imprint is published by Springer Nature
The registered company is Springer International Publishing AG
The registered company address is: Gewerbestrasse 11, 6330 Cham, Switzerland

All case law and rules citations contained in this work may be found in full at www.thelegaltechnologyguidebook.com

Preface

Lawyers' resistance to technology is legendary. Lawyers were reluctant to use phones to communicate with their clients when the phone was invented. It is a good bet that if you carefully check a storage closet in a law firm that has been around for a while you will find carbon paper, a Rolodex with phone numbers of many deceased people, and an ancient, yellowed pad used to make handwritten diary entries to bill the client. In a lawyer's house, the VCR is always flashing "12:00."

This book is an attempt to do something about that by communicating to lawyers the urgency of their understanding the technology around them and using it in their practice in a knowledgeable and efficient manner to serve their clients' interest. If we could ban one sentence in the English language, it would be: "you really do not have to understand the technology to get this." You most certainly do. Absolving lawyers from understanding the technology pertaining to the creation, maintenance, and storage of their clients' information is as dumb as telling medical interns that they do not have to understand "that X-ray stuff."

We have therefore written this book to communicate how to manage the technology that lawyers now have to use to collect, search, and analyze the information that their clients or their opponents have created. Whether involved in litigation or in a business transaction, the client's most precious asset may be the information it has created or collected. The days of telling the client to send to the lawyer a banker's box that has the information pertinent to a case or transaction are over. Now, the client may have a staggering amount of unorganized information that the lawyer may need but, unless the client has a superb information governance system, that information will be scattered all over a computer network without rhyme or reason. Learning how to find it, preserve it, collect it, and search for what is truly useful without bankrupting the client has become the lawyer's art and her necessary skill.

We begin with the lawyers' ethical rules that pertain to the required competence they must have and a discussion of other ethical rules that pertain to how a lawyer now must practice law because technology has so deeply affected the rules of the game. We could say that we do not intend to scare you but we do. You will see how lawyers can be seen punished severely if they persist in refusing to understanding the technology.

We then speak of the evidentiary rules and how these ancient rules are being used (or tortured) to deal with the revolution in how people communicate and create information that will be used at trial.

We then turn to the main course: How do you do it? We provide the most practical information we have as to how the available technology works, how to use it, and how to manage it so that the result is the best that can be achieved with the resources that are available. We explain how management techniques and processes can be used to check the results from the use of any available tool and how the lawyer should manage the process using the skills that come from an understanding of how information is managed in a digital world.

We hasten to point out that none of us can pretend to have had scientific training in data management. We learned everything you are reading in a very good school—the one of hard knocks. We hope that encourages you to learn what we have. It is as if we had been asked to write a book about roller-skating. We suppose that we could, but we prefer that you get up on the skates and we hope you enjoy the ride.

Washington, DC, USA   Judge John M. Facciola

Contents

Part I
Introduction to Technology Competence

Chapter 1
Introduction to Ethics and Technology

That there should be a distinct set of ethical rules pertaining to lawyers' use of technology seems odd at first glance. Ethics rules are written broadly by design and would seem to be able to apply across all practices; there are not, after all, distinct rules pertaining to the representing banks or trying malpractice cases. Yet, here, as elsewhere, the extraordinary changes that technological developments have made in the creation, processing, collection and preservation of electronically stored information have required that broad rules be refined and then applied to problems that no generation of lawyers have ever faced.

Imagine a law firm in 1975 and a visit from a client who has a legal problem or represents a corporation. The client discusses the problem with the lawyer who may take notes and, after the meeting, use those notes to dictate a memorandum to her file. The lawyer may then direct the client to collect whatever papers pertain to the controversy and deliver them to the lawyer who will probably makes copies of them for the lawyer's use, with the client possibly directed to maintain the originals. From that point on, lawyer and client will communicate by phone or letter and if by letter, a copy will be placed in the lawyer's and client's files. At the conclusion of the representation, the lawyer and client will make arrangements for the disposition of the copies of the client's files.

Now, imagine that same scenario in 2015. The client may well visit the lawyer but there the similarity ends. The lawyer has a computer and will record her notes or type a memorandum to her file on that computer. While the lawyer may not know it, the computer itself has created information about the lawyer's use of the computer and the work done and this information, the metadata, are accessible during any well-designed examination of the computer's workings. Similarly, every piece of electronically stored information that the client and its employees have ever created will have that same characteristic of metadata.

The client's files that the lawyer's needs are now entirely electronic and are not stored in a cold old banker's box but on a complicated computer network. The client's employees may be using that network for all communications, whether in writing, by voice or text or instant message. There may be no systematic direction,

© Springer International Publishing AG 2017
K. Williams et al., *The Legal Technology Guidebook*,
DOI 10.1007/978-3-319-54523-3_1

whether mechanical or human, as to what is to be kept and what disposed of, and, as is likely, there is no such system the client and lawyer are confronted with a gigantic data set, the vast majority of which has nothing to do with the matter. Somehow, someway someone is going to have to separate the few grams of wheat from all that chaff.

Once the information is gathered for the lawyer, the lawyer has to decide where and how to keep it. Its size may warrant a solution other than the transmittal of electronically stored information by email Or, the client may have already kept the information in cyberspace ("the Cloud") and the lawyer may wish to have the information accessible to her from that receptacle. Or, irrespective of how the client kept the information, the lawyer may find it cheaper and more efficient to maintain all of its clients' files in a cyberspace or cloud environment.

The lawyer's search within the client's files may produce problems of efficiency and cost if the lawyer, like the lawyer in 1975, intends to go through them item by item. Such a prospect may trouble a client who has to weigh cost against value. Technology may suggest to client and lawyer delegation to a third party to collect from the client's file what the lawyer truly needs and a particular vendor and its technology may provide an attractive means of doing what must be done.

The radical difference between the environment of the 1975 lawyer and of her 2015 colleague explains why more refined rules have had to be created to deal with the differences in those environments. Anything else would be to either pour new wine into old bottles, or mouthing a bromide ("A lawyer should be competent") that provides no actual guidance to the concerned lawyer.

Thus, there are new rules, or more accurately, new interpretations of traditional rules that impose such different obligations that they truly become, in themselves, new rules. This chapter is devoted to an explanation of those rules and their application.

A crucial word of warning: every ethics presentation begins with the warning that the Model Rules of Professional Conduct of the American Bar Association ("ABA") Formal Opinion are just that: models which state bars are free to accept, reject or modify. Therefore, a lawyer must consult her local bar rules of ethics before deciding on a course of action. That advice commands a lawyer's obedience particularly in this area. Lawyers will find that there are not minor differences between a given Model Rule of Professional Conduct and a local rule but that one directly contradicts the other. Lawyers fail to appreciate that at their peril. Indeed, if time permits, counsel is well advised to seek prospectively an opinion from a local bar before undertaking a particular course of action. Such rulings can provide certainty where, while the Model Rule of Professional Conduct provides guidance, the local rule does not and reasonable people could differ on the outcome. Given an area so full of dramatic technological change and where there already dramatically divergent approaches between the Model Rules of Professional Conduct and local bar opinions, counsel simply must use all necessary means to get certainty before embarking on a course of action. To fail to do so may be to walk through a minefield without a map armed instead with the hope that counsel is doing "the right thing."

Chapter 2
Technological Competence

The technological change lawyers have witnessed brings in its wake an obligation to know enough about that change and its consequences, if they can be reasonably anticipated, to have an impact upon the client's business and the representation to be provided.

Model Rule of Professional Conduct 1.1 requires that a lawyer provide competent representation to a client. "Competent representation" is said by the Rule to require "the legal knowledge, skill, thoroughness and preparation reasonably necessary for the representation." Id.

The Comment to the Rule indicates that whether the lawyer has that skill is a function of a comparison between the nature of the matter with the lawyer's training and experience in handling such matters and the feasibility of associating with a lawyer who has "established competence" in the area. Model Rule of Professional Responsibility 1.1, Comment 1: the clear inference of the Rule is that if the lawyer cannot provide "competent representation" she should decline the representation.

For present purposes, the most significant modification of the accompanying comments is the one generated by the American Bar Association Formal Opinion in 2012. Its president had urged the committee responsible for the Rules to bring them kicking and screaming into the technological world in which lawyers live. The product was a modification of Comment 8, "Maintaining Competence," which required a lawyer to "keep abreast of changes in the law and its practice" by adding the words "*including the benefits and risks associated with relevant technology.*"

A moment's thought indicates obvious instances where lawyers' behavior can and will be tested by this standard. Take a discovery dispute where lawyer A demands a certain production of electronically stored information and lawyer B tells the judge, considering a motion to compel, that the production would be difficult and cost hundreds of thousands of dollars. If it turns out that that lawyer B's assertion is false and uttered by him without any effort to understand how available technology could have lessened those costs substantially and that knowledge was easily accessible and well known by other lawyers who litigate eDiscovery cases, lawyer B may well have violated this rule.

© Springer International Publishing AG 2017
K. Williams et al., *The Legal Technology Guidebook*,
DOI 10.1007/978-3-319-54523-3_2

Take another lawyer who uses a computer or tablet to work while not in his office. He does not encrypt the contents of the device he uses, and uses as his password, giving access to the contents, "1234." When he does log on, he uses a public network, available "for free" in a retail store or an airport, to transmit highly confidential and privileged material back to his office. If his computer or tablet is stolen, or if his transmittal is hacked and diverted to a third party, breaching the privilege, and exposing his client's confidential trade secrets to its competitors, he may well have violated his ethical responsibility to be competent.

Recently, the Standing Committee on Professional Responsibility and Conduct issued Formal Opinion No. 2015-193 (hereafter, "Op.") which gives the most detailed and precise advice to date on the specific ethical obligations that are imposed upon a lawyer who is involved in what the Committee calls "the handling of discovery of electronically stored information." Op. at 1.

Presented to the Committee was a hypothetical where after an unsuccessful effort to resolve their differences concerning eDiscovery, counsel, at a judge's direction, meet again. One lawyer proposes a joint search of the other lawyer's client's network, based upon a jointly agreed search term list. This lawyer offers the other a "claw back" agreement which permits the client to retrieve ("claw back") any inadvertently produced electronically stored information that is protected by the attorney-client or work product privileges. Op. at 1. The attorney for the other side mistakenly believes that this agreement will permit his client to claw back anything it produces, whether or not privileged. He agrees with the proposal and the agreement is made a part of the judge's case management order.

This attorney exchanges his search terms with opposing counsel and they agree that both sets will be used.

The CEO of his client, known to be a large company with an information technology department, tells the lawyer that there is no other electronically stored information in its possession that has not been already provided to counsel. The lawyer accepts that representation, assuming that the client's IT department knows how to search a network and directs the client to make the network accessible to other side on the date agreed upon for the search, using the agreed upon search terms. The lawyer does nothing more; he never reviews what he has been given but instead assumes that the network search will not yield anything more than what the client has already given him. When the search is conducted, its results are provided him in electronic form but he does not review, apparently, because he is busy with other matters.

A week later, however, the lawyer gets a letter from opposing counsel accusing him of destroying evidence. When the attorney cannot open the electronic results of the network search, he hires an expert who accesses the data, conducts a forensic search, and tells the lawyer that:(1) potentially responsive electronically stored information has been routinely deleted from the client's computers as part of its normal document retention (or destruction) policy, resulting in gaps in the production; (2) due to the breadth of the agreed upon search terms, both privileged and irrelevant but highly proprietary information about the client's upcoming revolutionary product were provided to its chief competitor; (3) an IT professional, with litigation experience, would have recognized the overbreadth of the search and prevented the retrieval of the proprietary information. Id. at 2.

The Committee began its Opinion by explaining that, while lawyers are required to be competent, a mere failure to act competently does not trigger discipline, unless it was reckless, intentional or repeated. Nevertheless, the reality was that nearly every case potentially involves eDiscovery. Therefore, the ethical duty of competence requires the lawyer to assess at the commencement of every case what eDiscovery issues will arise. If they will, the lawyer must assess his own skills and, if he lacks them to associate or consult with someone who has the expertise he lacks.

The Committee then specified that lawyers handling eDiscovery must have the following skills.

(1) initially assess eDiscovery needs and issues, if any;
(2) implement/cause to implement appropriate ESI preservation procedures;
(3) analyze and understand a client's ESI systems and storage;
(4) advise the client on available options for collection and preservation of ESI;
(5) identify custodians of potentially relevant ESI;
(6) engage in competent and meaningful meet and confer with opposing counsel concerning an eDiscovery plan;
(7) perform data searches;
(8) collect responsive ESI in a manner that preserves the integrity of that ESI; and
(9) produce responsive non-privileged ESI in a recognized and appropriate manner
    Op. at 3–4.

The lawyer failed in obvious ways to meet these obligations. He began by failing to evaluate how eDiscovery would impact discovery in the case and his own capabilities to meet the consequential obligations that would be imposed upon him if there was to be eDiscovery. This attorney, however, made no assessment of either, and made the situation worse by not consulting with another attorney or eDiscovery expert prior to reaching the agreements he did. Similarly, he agreed to search terms without any understanding of how that could operate to permit the disclosure that occurred.

He also failed abysmally to instruct his client in how to conduct or supervise the search of the network. He did not pre-test the search terms or review the client's data before the search, seeming only to rely on his assumption that client would know what to do, and, in any event, that the claw back agreement would protect his client from the disclosure that occurred.

According to the Committee, all of this could and should have been avoided by the lawyer's consulting those who knew what he did not who could have informed him and steered him away from his improvident agreement and helped him to structure the search to avoid the consequences this search engendered.

Chapter 3
Outsourcing

Recall again that in law offices in 1975 there were only three kinds of people: partners, their secretaries and associates. The order is intentional. In 1975 associates were the lowest level of professional life in a law firm. There appeared, however, a new form of life, a paraprofessional. Neither associate nor secretary, the paraprofessional claimed she could do many of the duties being done by associate but at a much cheaper rate. Initially, it was a hard sell and paraprofessionals seemed, at times, to be chained to the Xerox machine. But, paraprofessionals proved their case, combining efficiency with cost saving much to the delight of clients.

Fast forward to 2017 and in the modern law firm the lawyers may start to feel like a minority. There are business managers, marketing personnel, technicians introducing technology and maintaining it, and a slew of non-lawyers who are doing most of the eDiscovery work. But, a tough question arises: if lawyers are obliged to be competent and serve their clients capably, can they ever delegate that responsibility?

The ethical rules pertaining to a lawyer's supervision of others are actually quite forgiving. Rule 5.1(a) of the Model Rules first indicates that a partner must make reasonable efforts to ensure that the firm has in effect measures giving reasonable assurance that all lawyers in the firm conform to the Rules of Professional Conduct. Similarly, partner or not, any lawyer who has supervisory authority over another lawyer also must make reasonable efforts to ensure that the other lawyer conforms with the Model Rules. Model Rules of Professional Conduct 5.1(b). But, the Model Rules are forgiving and excuse the failure of the inferior being cause for sanction of the superior unless the superior knows of the offending conduct at a time when its consequences can be avoided, or mitigated, but fails to take remedial action. Model Rules of Professional Conduct 5.1(2). The same exact principles are applied by the Model Rules to supervision of non-lawyers by Model Rule of Professional Conduct 5.3. Again, the supervising lawyer is obliged to make reasonable efforts to ensure that the behavior of the non-lawyer she supervises is compatible with the Rules.

© Springer International Publishing AG 2017
K. Williams et al., *The Legal Technology Guidebook*,
DOI 10.1007/978-3-319-54523-3_3

This is, of course, not the traditional standard that imputes, for example, tortious liability on the master for the negligent acts of the servant, so long as the acts were within the scope of the employment of the servant. Instead, the supervising lawyer is only responsible if: (a) she fails to take reasonable efforts to ensure that the inferior lawyer conforms to the rules, or (b) she knows of the unethical behavior when its consequences could be avoided or mitigated, but fails to take action to remedy the impropriety. Thus, the cynical would note that lawyers on behalf of their clients impose liability upon third parties for their employees' misbehavior so long as it is within the scope of their jobs but excuse themselves from the same liability.

eDiscovery has raised a new set of problems for the delegating lawyer. At the its inception, lawyers did the review of the data to be produced or that had been produced by opposing counsel themselves by delegating the responsibility to junior associates and billed the client at their ordinary rates. Given the data sets that the lawyers reviewed and those rates, the screams of anguish from their clients when the bills arrived must have been something to hear. There followed delegation of the review to contract attorneys who were hired at reduced rates for a specific job and did not qualify for the benefits afforded associates, lowering overhead and the clients' bills markedly. The next stage was complete offshore delegation where the data to be reviewed was done by foreigners (for want of a better word) at truly bargain basement rates. The final and present stage is the present one where the process of collection, review and storage of the data, is done by a vendor who charges a fixed rate. Finally, many more clients are taking the process in house, doing the collection and review with their own personnel and presenting litigating counsel, who is representing them, with a finished product.

Whatever process is chosen, serious and cognate ethical rules are engaged. First, if a lawyer is obliged to be competent, how does she fulfill that responsibility when the work is being done by a non-lawyer? Second, the lawyer is obliged by Model Rule 1.6(c) requires a lawyer to make reasonable efforts to prevent the inadvertent or unauthorized disclosure, or unauthorized disclosure of, or unauthorized access to, information relating to the representation of a client. How does a lawyer comply with this rule when the entire corpus of the data, privileged and not privileged, is made available to a non-lawyer who may be in Bangalore, India, and who is storing all of that data on a third party's server so that the data is said to be in the cloud?

Third, a lawyer may not represent two clients when their interests are adverse. Model Rule of Professional Responsibility 1.7(a). Does that obligation limit the ability of a vendor whom the lawyer has retained to render services to another lawyer when the lawyer could not because of a conflict of interest? Finally, a lawyer may not aid another person in engaging in the practice of law "in violation of the regulation of the legal profession in that jurisdiction. Model Rule of Professional Conduct 5.5(a).

In a 2008 opinion, the Ethics Committee of the American Bar Association provided comprehensive guidance to the outsourcing lawyer in these situations. American Bar Association Opinion 08-451 (2008).

First, as is obvious, the lawyer fulfills her responsibility to her client when the third party that is retained provides competent representation. But, the outsourcing must do more than await the results. Instead, she must investigate the background and references of the persons providing the services: (1) what are their hiring practices? (2) are the persons employed investigated before they are employed? (3) how is the quality of the services provided evaluated? (4) if lawyers, what is the system of legal education and how are disciplinary rules as to lawyers' behavior enforced? (5) are the lawyers and non-lawyers instructed as to the American legal system and its requirements as to, for example, production under American discovery rules and legal ethics pertaining to attorney-client privilege or work product protection?

Second, with reference to the confidentiality of the data sent to the provider, is it subject to seizure in the foreign country, despite claims of ownership by the client? How responsive will the legal system be in resolving any legal dispute that erupts between the outsourcing lawyer and the person or company retained? What efforts does the third party have in place to protect the confidentiality of the data and are they consistent with the outsourcing lawyer's responsibility to take reasonable efforts to protect the client?

The lawyers should get honest and verifiable answers to this question and the Opinion cautions that the lawyer may have to visit the foreign country where the vendor is to get the answers she needs.

According to this Opinion, the client must approve this delegation; an earlier opinion indicating that a client did not have to consent to a temporary lawyer performing a legal task on a client's behalf was distinguished. Furthermore, the fee charged by the vendor who provided the services may not be marked up. The client can only be charged what the vendor charged the lawyer plus overhead.

While this Opinion declined to indicate whether the retention of the vendor engaged the hiring lawyer in aiding the unauthorized practice of law, the D.C. Bar has not been so shy. In its opinion, a lawyer, who, for example, owns a company that provides eDiscovery services to lawyers had better be careful if that company provides legal services or legal advice that lawyers and lawyers are permitted to provide (DC Bar Ethics Opinion 362 (June, 2012)).

A common phenomenon is the lawyer's attempting to save the costs of storage by negotiating a contract with a vendor who provides storage on a server for the lawyer's files, including the productions generated in litigation, where there has been produced or generated large quantities of electronically stored information. While the vendor may not be providing a service other than storage, the inevitable presence of the client's confidential information in these files engages the lawyer's responsibilities to make reasonable efforts to preserve the confidentiality of that information.

The ABA Commission on Ethics 20/20 Working Group on the Implications of New Technologies, published an "Issues Paper Concerning Client Confidentiality and Lawyers' Use of Technology" (Sept. 20, 2010) and considered some of the concerns and ethical implications of using the cloud.

The Working Group found these potential confidentiality problems involved with "cloud computing:"

- Unauthorized access to client-confidential data by a vendor's employees, subcontractors, or by hackers or disgruntled or dishonest former employees who may still have password access to the vendor's files.
- Storage of client data in countries with fewer legal protections of electronically stored information than available, for example, in the United States or the European Union.
- Conversely, storage of the data in a country that has greater protections than the United States, rendering access to the data for discovery purposes more difficult (if not impossible) than it would be in the United States.
- Adequacy of back up processes by the vendor that will permit immediate production of the data if there is a technical failure (a server crash), the data becomes corrupted or destroyed by fire, flood, terrorist attacks, or if a hacker inserts a virus or other malware that infects the data, potentially destroying it.
- Unclear policies regarding data ownership, and preventing data being seized by a government or third party creditor with a claim against the vendor.
- Maintaining access to the stored data if the vendor: (a) is fired (b) goes out of business, or files for bankruptcy (c) suffers a temporary interruption of its business.
- Unclear polices as to notice of: (a) the lawyer (b) the client (c) third parties when there is a breach of security. What law applies and who will give whom what notice of the breach?
- Insufficient encryption of the client's data that could have prevented the breach, but was not used.
- Procedures for notice of lawyer and client from vendor when vendor is served with a subpoena by a government or a third party. Will vendor resist the demand until the lawyer and client can be notified and then decide whether the client, aided by the lawyer, should intervene and resist the demand.
- Will vendor have unequivocal obligation to resist the demand, or can it surrender the data without consequences?
- Policies as to verified destruction of the data when the client and lawyer determine it is no longer needed and should be destroyed.
- Policies as to transfer of the data when client changes lawyers, or when client and lawyer change vendors.[1]

State bars have spoken to this issue and emphasized how crucial it is that lawyers carefully review their clients' contract with vendors that are often adhesive and presented on a take it or leave it basis. From a valuable opinion by the Pennsylvania Bar[2] can be derived the following checklist for counsel's use when negotiating with a vendor who will provide storage of a client's or lawyer's files:

---

[1]Penn. Bar Assoc. Comm. On Legal Ethics and Professional Responsibility, Formal Op. 200–2011 at 13.
[2]Id. at 13–14.

Integrity of original data	—Backup and restoration if lost
Breach	—Firewall and refusal to disclose confidential information to the unauthorized without client's permission —Audit to monitor who is accessing data —Technology that can withstand breach: penetration testing
Confidentiality	—Written agreement to enforce obligation to provide security —Verify identity of individuals to whom attorney provides confidential information —Limit to need to know —Backups —Encryption —Specify how client information is to be handled —Plan to address security breach; who is to be notified, including client and law enforcement
Data Ownership	—Provider does not own the data, lawyer and client do
Review rights when subpoena and search warrant are served	—As stated
Audit rights	—Lawyer can audit provider's security procedures or to get copies of audits performed by others
Host server location	—Data only with geographical area; if elsewhere have to consider impact of laws of forum where stored
Termination	—How will data be retrieved when vendor goes out of business or has some break in continuity of service
Training	—Of all lawyers who use it
Compliance by users with data protection measures	—Lawyers must comply with security measures, such as strong passwords
Lawyer's due diligence	—Investigate vendor's: —Security —Recovery —Back up procedures —Safeguards against natural disasters —Where are the servers —History, funding and reputation —Policies for data retrieval upon termination and charges that are due —How will it comply with litigation hold
Data format	—In non-proprietary format
Retrieval of data when need offline	—How retrieve for the deposition
Internet connectivity	—What to do if Internet goes down
Credits for time offline	—If not 100% access, money credits for all time system is done

There would be some who would say that lawyers and their clients have so little bargaining power with vendors that insisting on all these terms is a pipe dream. Nevertheless, too many clients who accepted what they are given have on closer inspection found, for example, that liability in the event of a loss of data is $1.

Lawyers have to appreciate just how costly such surprises will be. Shopping might help. Lawyers have found that the smaller the cloud service, the greater the hope that terms can be truly negotiated. It is certainly worth the effort. Acceptance of a contract of adhesion that contains none of these terms and excuses the cloud service from any but the most minimal obligations may well be catastrophic when something goes wrong.

Chapter 4
Receipt of Privileged Information

Imagine a lawyer who receives an email from opposing counsel with an attachment, which the sending lawyer indicates, is about a certain topic. The other lawyer opens the attachment and is not about that topic at all. It is about some other topic and on the top of the attachment are the words "Attorney Client Communication." What must the second lawyer do?

Not much, actually, Model Rule of Professional Responsibility 4.4(b) indicates that, if the receiving lawyer knows or should know that the attachment was inadvertently sent he must notify the transmitting lawyer. That's it. There is no other obligation and certainly no obligation not to use it or transmit to someone else who asks to see it. Accord: D.C. Ethics Opinion 256 (while receiving lawyer must notify sender of inadvertence, there is no impediment to receiving lawyer using it in the next round of negotiations with the transmitting lawyer).

Electronically stored information raises a cognate issue. Metadata, or data about data, may be produced by the programming of a lawyer's computer or, in application software such as an Excel spreadsheet, there may be information imbedded in the document that is crucial to understanding the document but may not be apparent in the form in which the spreadsheet is produced. Thus, if column c shows the result of adding columns a and b, one would have to look at the underlying formula in column c to understand the calculation because the calculation is not shown when the spreadsheet is produced even in its native format.

Additionally, most word processing programs, such as Word or Pages, permit the user to track and see the changes that have been to the document as it has been edited or revised. Can a lawyer look at this underlying data if it is not immediately apparent? Unfortunately, there is no clear answer and the authorities are in utter disagreement.

First, the American Bar Association Formal Opinion 06-442 (2006) runs away from the problem. It says that Model Rule of Professional Responsibility 4.4 which, as shown above, deals with the lawyer's receipt of apparently privileged information has nothing to do with the metadata problem and that's that.

© Springer International Publishing AG 2017
K. Williams et al., *The Legal Technology Guidebook*,
DOI 10.1007/978-3-319-54523-3_4

Nature abhorring a vacuum, the states have stepped and made a hash of it, so that behavior that is permitted in one state is absolutely condemned in another. New York has a flat prohibition while the District of Columbia requires the receiving lawyer to notify the sender that she has received it only if she has actual knowledge that it was inadvertently sent. If she does not have such knowledge, she must actually be obliged to use it to fulfill her responsibility to represent her client competently. Thus, as just indicated, behavior that is condemned in New York may be required behavior in the District of Columbia, at least where the lawyer lacks actual knowledge of the inadvertence. It would be hard to imagine less careful guidance to the poor lawyer who practices in New York and Washington D.C. and where her ethical responsibilities may be an absurd function of where she opens a file.

This discrepancy may be explained by a very poor use of the word metadata, which, contrary to some of these opinions, does not have a universal meaning.

Take the first problem, looking at the calculations in an Excel spreadsheet. It is, of course, absurd to suggest that it is unethical to look at the underlying formula that yields the result in a certain cell. Insofar as the opinions barring metadata would bar looking at the formulas in a spreadsheet, they can't be right. That conclusion is irrational.

Second, take the track changes problem. Suppose it is a letter making a settlement offer of $100,000. Track changes shows, however, that it was originally $50,000. By knowing that the receiving lawyer knows that his opponent is bluffing and would take $50,000. She has now gained a crucial advantage in the next round of negotiations and that advantage was gained by her looking at the "track changes".

Third, a lawyer receives a massive production of electronically stored information that was supposed to be produced in TIFF but some were produced in native. A vendor does some sampling for her and finds several documents that have the heading "Attorney-Client Communication." The lawyer looks at the documents and they are innocuous transmittal letters from the attorney to the client, sending copies of documents filed in the public docket. She therefore concludes that her opponent had decided not to bother to claim the privilege and that the transmittal of the document and its metadata was purposeful and not inadvertent. She therefore tells the vendor that there is no impediment she can see to examining the metadata in these documents. She tells the vendor to find all the documents that bears the "Attorney-Client" heading. She reviews them and finds that, while most are innocuous, there are a handful, also in native format that might have been sent inadvertently in native so she holds off having her vendor find and catalogue the metadata.

She calls the lawyer who now claims that everything that bears the heading "Attorney—Client communication" be returned to him, that she return all the metadata she has collected from the documents, and that he will report her to the disciplinary authorities.

These examples show how the universal use of the word "metadata" can be so confounding. "Track changes" is an entirely different form of metadata than the

Excel spreadsheet and both of them are in turn different from the production. Certainly, in the "track changes" situation there is reason to believe that the transmittal of the data was inadvertent because a lawyer knows that any other lawyer would never show his negotiating "cards" to opposing counsel. On the other hand, in the large production situation, the lawyer made a responsible decision as to obviously innocuous documents. She was then forced to guess as to some of the documents where transmitting counsel's intentions were more ambiguous. The only way she could resolve her problem was to look at the documents but, once she did, she learned of their contents and, short of a lobotomy, she cannot be expected to forget what she saw, including the supposedly inadvertently sent metadata. Despite her good faith, she may still have to run the risk that a disciplinary committee will find that she should have realized that the transmittal of the documents in native was nor inadvertent.

It is maddening that all of this trouble is caused by the transmitting lawyer's failure to eliminate track changes before he transmitted the letter and by his failure to make sure the production was not in native with accompanying and unexpurgated metadata. Yet, the universal use of the word metadata obliterates the radical differences in the situations lawyers may confront and shifts the burden from the blundering lawyer to the receiving lawyer who has the unhappy task of divining the blunderer's intentions from behavior that is uncertain and ambiguous. One really should not risk a law license because one could not read another lawyer's mind and then failing to be your brother lawyer's keeper.

Perhaps, the American Bar Association Formal Opinion got it right in the first place. There is no ethical rule that can be construed to apply to what is called "mining metadata", and until one is adopted, the lawyer so incompetent that she does not know how to prevent the transmittal of metadata should have no one to blame but herself when something goes wrong.

Part II
The History of Technology Competence

Chapter 5
What Is Technology?

There is debate whether the lawyer's ethical obligation to maintain technological competence derives from, or is merely clarified by, Comment 8 of the Model Rules of Professional Conduct ("MRPC"), Rule 1.1. The fact is for most attorneys, however, where the obligation initially arose does not matter.

Comment 8 reads:

> To maintain the requisite knowledge and skill, a lawyer should *keep abreast of changes in the law and its practice, including the benefits and risks associated with relevant technology,* engage in continuing study and education and comply with all continuing legal education requirements to which the lawyer is subject. [Emphasis added]

What matters for most attorneys is simply trying to understand and meet this obligation at its most basic level.

The relevant inquiry in meeting this obligation is not understanding *why* lawyers must keep abreast of "relevant technology" to maintain the requisite knowledge and skill. The challenge in meeting our Rule 8 obligations is understanding *what* exactly does "relevant technology" mean?

Let's take a step back.

Technology is everywhere. Observing its advancement and growing power is the shared experience of our modern lives. We imbue technology with our hopes in everything from resolving our greatest human yearnings (is there life on other planets?) to our most mundane, quasi curiosities (how many steps did I take today?)

We use technology everyday. Its terms roll off our tongues with confident command. Folks of every income and age, culture and education use technology. Its use has pervaded modern society with widespread, shared understanding.

From our cell phones, to MRIs of our bodies, to GPS in our cars, we often don't go a moment, let alone a day without being reminded of the reach of technology into our lives.

© Springer International Publishing AG 2017
K. Williams et al., *The Legal Technology Guidebook*,
DOI 10.1007/978-3-319-54523-3_5

5.1   What IS Technology?

But, what is technology. The lawyer seeking to meet their Comment 8 ethical obligations will not find the answer to this question in the MRPC themselves. While Rule 1.0 specifically defines certain "Terminology" found in the MRPC, no mention is made of the term "technology".

Looking outward for a common-use definition, it is not surprising to find consulting a dozen dictionaries yields three dozen definitions. But, no court in the land will conduct an etymological analysis when assessing a lawyer's obligations under these rules. So, perhaps, it is more useful to understand technology as any tool, or tools, that enhances our ability to perform physical or intellectual tasks. In 3000 BCE, papyrus scrolls were the height of technological advancement. In 1950, it was the telex machine. Today, it's digital currency, dancing holograms, and driverless cars.

The most important thing to understand about the prescription of Comment 8 is, therefore, that as technology evolves, so too must lawyers' understanding of how technology impacts their practice, and their clients' personal and professional lives.

5.2   Identifying Technology Risks

As we consider this evolution, it is helpful to look at some of the issues surrounding technology with which courts and counsel have wrestled from an ethical standpoint.

We can see where new technological tools created fundamental shifts in how legal tasks are performed, and how the ABA then conceived of, or applied, rules governing the related behavior.

Technology based issues are not limited to developments within the walls of law firms or law departments. Issues may arise externally with clients and other outside sources.

By the mid 1990s, email had become a standard business tool in the working world. An ability to communicate instantly, around the world, for free (or, at low cost) with anyone who choses to give themselves an email address, brought forth a near overnight change in how most people, in personal and business practices alike, chose to communicate. But lawyers had an ethical conundrum. Whether or not sending email messages through well-known providers that were "unencrypted" constituted a violation of their duties of confidentiality and privilege.

Even where email was password protected, the issue was whether an electronic message passing through multiple servers, and which could, theoretically, be accessed by unintended third parties, was a violation of the lawyer's ethical requirement to keep client communications confidential. After years of debate, in 1999 the ABA's Standing Committee on Ethics and Professional Responsibility weighed in, finding that the reasonable expectation of privacy when

Communicating Over Unencrypted Email was enough to meet the lawyer's Model Rule 1.6(a) prohibition against revealing confidential client information.[1]

Relatedly, the ABA's Standing Committee on Ethics and Professional Responsibility has issued Formal Ethics Opinions on Lawyer Websites [Do they create an attorney client relationship under MRPC Rule 1.18? (answer: sometimes)];[2] Judge's Social Media use [Does using social media violate Code of Judicial Conduct Rule 2.9(A) prohibition on ex parte communication? (answer: no)];[3] and Outsourcing (including via the internet to non-US entities) [Does outsourcing violate MRPC Rule 1.1 competency obligations? (answer: if properly supervised, no)].[4]

Once the tool is proven that it is here to stay, the ABA will weigh in. What does this mean for the intrepid early adopter? Let's briefly recollect the text of Comment 8: [t]o maintain the requisite knowledge and skill, a lawyer should keep abreast of changes in the law and its practice, including the benefits and risks associated with relevant technology.

When dealing with relatively novel technologies, lawyers, and their support staff or contractors, should take this to mean that they must establish a familiarity with the use *and use case* of the technology at least adequate to understand and mitigate any risks—and to explain such factors and mitigating actions to an inquiring client, court, or opposing counsel.

Fundamental Shifts in Performing Legal Tasks

Currently, there are hotbeds of developing use that seem ripe for treatment by the ABA, or other state bar ethics committees. These are instances where the, increasingly widespread (if not ubiquitous) use of a new tool or software makes one lawyering task easier, but also introduces new complexity around security, confidentiality and privilege.

BYOD

Activities intrinsic to modern law firm practice are some of the most prevalent and pressing issues. A major topic of the last half-decade has been the evolution of the "Bring Your Own Device" phenomenon ("BYOD").

The days of a lawyer showing up to the office to work, diligently punching out their daily quotient on a single desktop computer and then leaving in the evening, not to think about work until the next day, are long gone. The accessibility of technologies of various types, and the demands of global clients, has erased the distinction between office and home. With this has come the disappearance of "work" computers and "personal" devices. As we are able to work from personal laptops, smartphones and

---

[1] ABA Op. 99–413.
[2] ABA Op. 10–457.
[3] ABA Formal Opinion 462.
[4] ABA Formal Opinion 08–451.

tablets, we often find that we DO work from those devices (and, also, that we must, which is a cultural shift for discussion at a later date).

What is clear is that adoption of BYOD has, in some respects, made work easier. Their use has also given rise to new concerns.

Issues Addressed by BYOD: cost, familiarity, ease of use and transport.

Issues Created: ownership, data privacy, exposure to networks, control (if lost or stolen).

Cloud Storage

Another major issue that has developed is around where documents, and other data, should be stored. Documents that were once created on the individual workstations of attorneys, or their support staff, began, by the early 2000s, being shared on local servers within law firms or corporations themselves. This allowed for easier sharing of work product, templates, and greater version control. While, local servers kept such information behind a firewall (in theory) there soon came a push to move files to the cloud.

While cloud storage provides more manageable costs structures, more robust data security measures than most law firms can take on, and on demand scalability, in many instances, it also introduces its own uncertainties. For example, where does the data reside? Who controls the data? These are questions that may be of critical importance for lawyers engaged in cross-border discovery, or information governance.

Issues Addressed by Cloud Storage: costs, access, knowledge management security.

Issues Created: security (of a different kind), confidentiality, jurisdictional questions.

5.3 Identifying Technology Benefits

It is a common refrain among attorneys who have practiced for many years quite ably without the benefit of advanced technology, to lament the "intrusion" of technology into the profession. With the evolution of an affirmative obligation to keep abreast of technological change it can appear to many (especially those who may have encountered BYOD, Data Storage, or related problems), that adopting new technologies into their practice is the *cause* of, not the solution to, their problems.

There are more than a few attorneys who would simply avoid these problems by maintaining the internal systems and strategies they have used for years. But, frustrating as it may be to the old guard, or the slow adapters, standing still while the world evolves is not an option. The world is evolving. So must the lawyer.

Those who focus only on the potential problems posed by technology, risk missing the potential for technology to positively impact their legal practice. The benefits are quite profound. Not only can new technology help lawyers deliver more cost effective, and consistent services to their clients, many new tools, from eBilling systems to practice management platforms, help lawyers run their businesses with greater efficiency and cost-predictability.

And, of course, there are plenty of people willing, for a small fee, to lend a hand.

For every risk or inefficiency that lawyers encounter, there seems to be a vendor offering a hardware or software solution. Legal technology is a $300B per annum market. There are over 1000 self-identified legal technology vendors. This does not include the many software, and other tools, available for attorneys that are not "legal" per se. Email, CRM (Customer Relationship Management) tools, word processors, scheduling, billing, among many other "general" business tools all come to mind.

The legal technology market, new in many respects, has traits associated with all developing markets: new entrants, mergers, trends, new product classes, etc. Many of these products are tremendously powerful and capable. But how do we proceed with simply meeting the obligation to keep abreast of their benefits?

One strategy is to categorize the tools in a broad manner that helps us understand their capabilities. In Appendix F, we have provided a taxonomy that organizes the universe of relevant technologies based on their primary function:

1. Practice Management (virtual law offices, accounting timekeeping document management, etc.)
2. Litigation (conflict checking, eDiscovery, data retrieval)
3. Security/Storage/Backups (anti-virus, firewalls archiving)
4. Patents (patent searching)
5. Information Decision Tools (data analytics flowcharts, decision trees)
6. General Office Tools (word processors, spreadsheets)
7. Collaborative Tools (content management, file sharing)
8. Big Data (artificial intelligence)
9. Legal Services (outsourcing, alternative legal practices).

The list is by no means exhaustive, but it is a useful tool to begin understanding and categorizing the landscape. As you begin, or continue, your own exploration, you will likely add, or modify, classifications.

5.4 Conclusion

For the creative and ambitious lawyer looking to leverage technology for the better, the benefits of thoughtful adoption will be profound.

While it is impossible to predict the response that various oversight bodies may have, technology continues, at an increasing pace, to penetrate legal practice environments. We can certainly expect that such bodies will continue to issue opinions and guidance Since the promulgation of such rules is, necessarily, behind the curve of actual adoption, it is, however, critical that attorneys continue to be mindful of how their use of new technologies is impacted by their ethical obligations of confidentiality, privilege, diligence and cost constraint, as set forth in the MRPC, and elsewhere.

Chapter 6
How Did We Get Here?

It is still hard to believe that in 1997 the cost of restoring certain back up tapes in *McPeek v. Ashcroft*, 202 F.R.D. 31 (D.D.C. 2001) that so troubled the court and the parties was estimated to be $774 (i.e. 8 h @ $93 an hour). Lest one thinks that the judge and the parties had taken leave of their senses, it must be recalled that in a paper universe spending $700 on the production of paper was a substantial amount of money. 10¢ was the going rate per page for copying and in that world $700 was a lot of paper. It is impossible to exaggerate the difference from 2001 to today where parties to a lawsuit pay their vendors several times that for each gigabyte that is processed, collected and stored.

With a rise in costs, there has been an equally significant change in who does the work. In the early days of eDiscovery, law firms filled a room with associates who they billed at substantial hourly rates, and who were expected to look at each document produced or to be produced. This process may have been the occasion of increased risk of heart attacks when the client saw the bill. The process hardly resembles the present reality with larger firms either paying vendors to do the processing and collection, or creating their own eDiscovery centers were contract lawyers, paraprofessionals, and persons skilled in analytics now do the work. This development is, of course, the product of the client's demand for lower or fixed fees per representation, and more intense supervision of how the work is being done. Indeed, larger clients are creating their own substantial in house capability to do the necessary collection, processing, and searching.

Thus, the world of the lawyer reviewing a box of documents is a quaint recollection of a time that has gone by. As the cloud replaces the banker's box, we will see the more aggressive use of artificial intelligence and machine learning to reduce even more the high price of human beings doing the work. Obviously, there is a new demand that lawyers secure quickly the technological competence to understand how data analytics and artificial intelligence work so that they can make the correct judgment of how to fulfill their obligation to provide competent representation when the rest of the client's business is being focused on how to save money and time using similar "machine learning" tools. It is unreasonable for lawyers to expect that

© Springer International Publishing AG 2017
K. Williams et al., *The Legal Technology Guidebook*,
DOI 10.1007/978-3-319-54523-3_6

their work with their clients will be partitioned from the client's work with all its other providers of goods and services, which is based on modern tools of data analytics.

Finally, lawyers will have no choice but to gain this competence. The capacity to store information is reaching nearly ridiculous heights, with a terabyte hard drive selling for $54, and with talk of using new technology to collect information in a way that will permit a library of data to be enclosed in a device the size of a sugar cube. Those who say this is impossible had better remember what was said as to other forms of technology, once thought impossible, that have become ordinary parts of daily life.

Additionally, unlike the back up tapes in *McPeek*, that were to be used only in an emergency, data has become a profit center in itself so that the client may be reluctant to dispose of it in the face of the argument that it can be used in the client's business or sold for a profit. If the client is a believer in information governance there is some hope that only what is valuable will be kept. If not, there will be the maddening retention of everything without rhyme or reason on storage devices or in the cloud where the decreasing cost does little to encourage a disciplined approach to data management.

The modern lawyer, therefore, confronts the obvious obligation to stay current with ever developing technologies, and in particular, artificial intelligence, to have any hope of competently representing clients whose needs and demands seem, at times, to be changing on a daily basis. To live in that world without a fundamental understanding of how information is kept, processed, maintained and searched is to sail a boat without a compass, let alone a GPS.

Chapter 7
Technology: Then and Now

With the 'a' and then the ring around it.
What is internet anyway?
What, do you write to it like mail?
I have no desire to be a part of the Internet.
                    Bryant Gumbel and Katie Couric (NBC 1994)

How did we understand technology twenty years ago? How do we understand technology now? Where will we be in another twenty years? Will there be a video clip of us at which future generations laugh due to our current lack of comprehension? Will we still call it a "video clip?"

Just as news anchors Bryant Gumbel and Katie Couric discussed the wonders of the Internet in 1994; it is likely that we similarly fumble with our understanding of many current, and certainly future, technologies. In 1994, users were concerned with how the Internet actually worked. It was an unfamiliar thought that someone could almost instantaneously send a communication. Prior to the Internet, people had more control over communication. If someone called, they could choose not to answer the phone. If they shredded a letter, they knew it was gone from their possession.

With the Internet and advancement in technology over the past twenty years, this way of operation has drastically changed. If someone calls and you do not answer the phone, there will be a "missed call" in your call log, there could be a voicemail linked to your electronic-mail account (or, as we must refer to it, email) and there could even be an email that transcribes the voicemail that was left by the caller. This scenario would make it almost impossible to deny the existence of this call. Now, let us consider the letter, which we shall now analogize with an email (*because who sends letters anymore?*). If someone sends you an email, you cannot simply throw it away, shred it or burn it. Yes, you can press "delete," but what does that do? Even if you threw your phone or computer, on which you received the email, out the window, burned it or smashed it, there would likely still exist multiple copies of that email. The sender would have a copy; the email service provider would have a copy; and if sent over a corporate network, the email would likely be archived on a server and maintained for as long as archives are stored.

© Springer International Publishing AG 2017
K. Williams et al., *The Legal Technology Guidebook*,
DOI 10.1007/978-3-319-54523-3_7

So how do we control this continuous flow of information? How do we dictate what communications we want to receive, who can receive communications from us, and who has access to such communications? How do we choose not to answer the phone, or to throw out that letter? Perhaps the better question is: Can we?

Why is understanding technology important for attorneys? You may already consider yourself an expert in technology, and perhaps you are. For the majority, however, that is likely not the case. Many may still be just as baffled as those from 1994. Others may have a false sense of comprehension. Yes, you may know that when you press send, your email travels over a network, landing in the electronic inbox of the intended recipient, oftentimes, within seconds. If you are really tech-savvy you may even understand how your email communicates with the server to be sent over the network. Generally speaking, this may be sufficient in your day-to-day affairs. As an attorney, however, you should have a more specific understanding.

Let us start with the basics.

7.1   The Network

In its simplest form, a network is a connection of two or more computers, in which one computer can communicate with another computer. Most commonly, these computers can connect to a network via cables, satellites or wirelessly. Think about your home. If you have a router and multiple devices that connect to that router, then you have a network. You can connect to this network directly with a cable, or you can connect wirelessly. This network grows as you add more devices. A given network, however, is designed to handle only a set bandwidth (a given capacity). If you exceed that bandwidth, then the network will operate at suboptimal capacity. As such, the network you have at home is likely not the same network you have at work, which would require a greater level of bandwidth.

There are two types of networks: (1) Local Area Network (LAN) or (2) Wide Area Network (WAN). A LAN describes, as you may have guessed, a local network with limited range, such as one found in your home, school or other relatively small area or building. The bandwidth of a LAN is proportionate to its reach, and as such, can be overwhelmed with too many devices. A WAN connects a larger geographic area, such as a state, a country or the world. The Internet, for example, is considered the world's largest network, allowing millions of computers to connect at once. A WAN, as you can imagine, is a complex sequence of smaller networks, linked to form a wider network to permit global communication, almost instantly.[1]

---

[1]See, Appendix A. Case Study: Network Security.

7.2   **Backdoor Access**

If, when you hear "backdoor access," you think of a thief, in the night, dressed in all black, carrying a big duffle bag through the back entrance of a building, then you are *almost* correct. You are in the right ballpark, but way out in left field. A backdoor in a computer creates an entry point into the system whereby a user can get access by bypassing the standard authentication requirements. Oftentimes, backdoor access is developed into a system for troubleshooting purposes. For example, the backdoor can provide access to support personnel when you contact a company's helpdesk for technical support. Although a benefit and convenience in some instances, the backdoor can also be a method for hackers to penetrate a system. Most backdoors are undetectable, and as such, hackers can access your systems without notice (now, you can see the similarity to our "thief in the night" comparison).[2]

7.3   **Historical Context**

As the legal profession begins to evolve beyond a profession that resembles what we know now, information, or more specifically the sharing of information, must become both functional and accepted. If the legal system is to become standardized, the need for widespread dissemination of information becomes paramount. The entire concept of "common law" being the law of the people is based on the concept of rulings based on precedent and the reliance on the judgments made in the past. Without knowing what the opinions of those learned judges are, lawyers in that system would have nothing upon which to base their arguments. Likewise in the "civil law" context, the law of the land must be written down and civil and criminal codes must be available for all to know and apply. Perhaps Civil Law more than Common Law relied on the advancement of knowledge sharing and technology at the time, but both have needed advanced printing and publishing techniques.

Consider that the first encyclopedias or compendiums of human knowledge were published in the early 1700s. If you were to know that the first legal reporters were available at approximately the same time, you might be surprised at how quickly the legal profession of the 18th century was willing to adopt technology. Case law was almost immediately organized by court and date so that successive lawyers could understand what the law in their jurisdiction was and what has been successful or not in arguing similar facts and laws. As courts became more complex and the rulings became more voluminous, so did the methods in which these reporters were published and disseminated through the legal profession. By the middle of the 19th century, US reporters were being published on large scales and were known to be accurate and widely available.

---

[2]See, Appendix B. Case Study: Backdoor Access.

We tend to think that the law is slow to adopt technological advances, and that may be true of modern devices and software, but when there is a need, a true requirement to disseminate and share information, lawyers have led the way over the last centuries. It is only in the last century, the digital age, that the legal profession perhaps hasn't seen the *need* for technology.

Philosopher and Essayist George Santayana famously extolled, "Those who cannot remember the past are condemned to repeat it." Life of Reason, Volume I. What often gets overlooked with this famous quote is that he was discussing progress—in particular societal progress, not war or foreign relations, where the quote is often referenced.

Critical point in the trajectory of the existence of the modern lawyers: technology is outpacing the lawyer's ability to utilize and understand it. As we stand on the precipice looking forward into a world where lawyers must proactively educate themselves of the risks and benefits associated with technology, it is important to take a moment and look back, and understand some of the factors that led to lawyer's technological incompetence in the first place. We need to understand where we went wrong.

First, there is a critical difference between "tech savviness" and "tech competence". Lawyers often wear two hats. They are business persons that must operate a profitable entity that meets the needs of their clients They also part of a much larger legal system, upon which there are rigid rules and ethical guidelines with which they must comply to maintain the ability to practice.

Up until this point in history, the ability to understand and harness the power of technology, within the context of daily lawyer life, provided an advantage (in theory) to the business aspect of practicing law. While clients are generally less concerned with the ability of their lawyers to be masters of technology, a general tech savviness has been seen as useful feather in the cap, and possibly a competitive business advantage.

What changes with comment 8 is that technological competence is no longer an issue for when the lawyer is wearing the client facing, business operation hat, but when they are wearing the hat of an advocate involved in the legal system facing aspect of the business. No longer does being tech competent allow lawyers to win the game of providing legal services, it is a requirement to play the game at all.

Lawyers often are painted as failing to adapt because the tech-barren methods that they use allow them to remain inefficient. Inefficiency, and its relation to the billable hour method of lawyer payment, being a critical element of maintaining legal practice as a professional avenue toward personal wealth. To the extent that maintaining inefficiency has been part of the failure to evolve, lawyers should bear criticism.

Hubris, the failure to compensate associates for their time spent evolving their practice (and only recognizing the value of hours billed), lawyer's failure to develop proactive business models and only reacting when clients present novel issues are all factors in creating these problems. An industry that fails to align the objectives of the clients with the objectives of the practitioners will quickly open itself up to competitive disruption. It may be too late for some lawyers for whom technology may supplant their practice anyway.

But it is not all the lawyers fault.

There are plenty of systematic issues that have dissuaded if not flatly rejected legal tech innovation. The US Legal system is uniquely complicated. Everything, and everyone is interrelated, and must do things in uniform manners according to uniform rules, which are often undermined or contradicted by varying State Bars, within law firms or in-house law departments, then with the courts systems, with opposing parties, and with clients. The whole industry is a giant web that must maintain quasi-uniformity. When deciding on the rules of how to play the game, the whole industry has to move forward together, and yet, this is often not possible.

Lawyers are also not incentivized to innovate. Yes, some technology represents opportunities to compete with other firms vis-a-vis creating better client experiences, many legal technology products allow for better/more efficient practice itself. But lawyers don't have the authority to use new tools. In fact using tools in a way to drive efficiency may itself be a potential ethics violation.

UPL regulations force extreme conservatism. Regulatory and Bench concerns exist as well. Some jurisdictions the courts don't even allow for electronic filing, how will they react when they rely heavily on a predictive coding tool for their document reviews? When existing in an ancient infrastructure, bold moves can collapse the process as much as they can advance it. Lawyers can only adopt innovative strategies so much on their own. All law is local, and if the judge they are in front of takes issue, the potential ramifications are far worse than the benefits. Simple risk/reward.

Another factor that needs to be understood is that innovation is happening exponentially, and yet adaptations are made linearly. The time it takes for the ABA and State Bar Associations to recognize a new trend and weigh in on it, is too long. Progress can't be tamed. The disconnect between the pace of innovation and how lawyers attempt to control it is flatly unsustainable.

Forcing lawyers to play within the rules of the system, while innovation is running amok, simply means lawyers will be forced to choose between the tools or the ABA's blessing. If we have learned anything from the Uberification of America, we are one major legal tech company away from the ABA not having any power at all. If society sees a better way of providing legal services, they are not going to wait for the ABA to bless it. Forcing the lawyer to choose between new tech and meeting their ethical guidelines becomes an impossible choice.

In a time like this we like to be able to look to our local bar associations or law schools for leadership. While there are exceptions, and pockets of hope, the few lawyers who are brave enough to look at the landscape of leadership must wonder, is it too late?

7.4   Moore's Law

In 1965 Gordon Moore, a co-founder of Intel, stated that the number of transistors per square inch on integrated circuits doubles every year and this will continue for the foreseeable future. The capability of adding more and more information to smaller and smaller real estate on the circuit has led to a staggering advancement in the capabilities of computers and information devices. Access to data and information in the last quarter of the twentieth century, and the beginning of the twenty-first century, has become pedestrian, meaning what used to be reserved for large machines or manual labor can now be done using phones and watches. Technology's rapid advancement has placed the human universe of knowledge at the fingertips of everyone willing (and able) to purchase a mobile device and pay for airtime.

This poses a particular problem for the modern lawyer. No longer is the lawyer alone able to access information using specialized technology such as case reporters and exclusive law libraries. The modern lawyer faces the challenge that her client has as much access to information if not more than she does and may have more advanced technology available to them as well. No area of the law at this time is immune from the reach of technology in everyday life. Moore's prediction that infinite amounts of data will be able to be stored on tiny devices has created a world where knowledge is shared instantaneously; where images and locations are published almost without the cognizance of those being tracked; and where recorded data of every aspect of our lives is stored and kept without limitation.

How can the modern lawyer keep up? How can she know where materials important to her case reside? How can she know she has *ALL* the information she needs to appropriately represent her client?

7.5   Where Are We Now

It is hard to say exactly what the breaking point was. In the US, much of the gravitas that comes with being a member of the bar comes from working within a common law system that proudly builds off of case law precedent that is every bit a binding law as the statute that was passed last year.

The legal system seemed to be a bastion of insularity. Unlike other industries that operate within the realm of the economy in general, lawyers have continued to exist within their own universe. Creating their own rules, and holding firm that society's pace of change would have to bow to their blessing. But society changed too greatly. Technology became too pervasive.

It was only a matter of time. Somewhere in the last two decade lawyer's inability to integrate technology into their practice went from quaint, to curious, to frustrating, to confusing. In the last decade the tech divide has moved to unsustainable and, finally, unethical.

There is no single incident of tech incompetence that brought the ABA House of Delegates to rush to action during the middle of the summer of 2012, finally amending comment 8 to MPRC 1.1, and making technology competence a requirement seems to have been more of a result of general trends.[3]

It should be noted that the Model Rules of Professional Conduct are just that– Model Rules. They only have binding effect when adopted by individual state bar associations. As of this date, 20* state bar associations have adopted the comment, sometimes with modified language.

Additionally, while not in the language of the comment itself, the key takeaway is that that lawyer's technological competence obligation is a standard of reasonableness. Lawyers are not obligated to become computer scientist overnight. They aren't expected to become technologists. Though where exactly the line is, is unclear and will develop over time.

Comment 8, for the first time, puts the affirmative duty upon attorneys to proactively keep abreast of technology. The lack of specifics within the comments, for the first time, forces to lawyer to find the means and information to maintain their basic ethical obligation to practice.

---

[3]Model Rule 1.1 States:

A lawyer shall provide competent representation to a client. Competent representation requires the legal knowledge, skill, thoroughness and preparation reasonably necessary for the representation.

Comment 8,

To maintain the requisite knowledge and skill, a lawyer should keep abreast of changes in the law and its practice, including the benefits and risks associated with relevant technology, engage in continuing study and education and comply with all continuing legal education requirements to which the lawyer is subject. (Emphasis added.)

Chapter 8
Benchmarking Technology Competence

8.1 Looking Beyond Legal

There are a number of process and certifications that have been established, over decades, for managing technologically complex business processes. As lawyers, and legal professionals, we do not have to invent the wheel in this space.

Depending on your practice environment, you will have to decide which tools and training offer the best ROI (as well, as which members of your organization should receive training, or participate in the required project teams.)

8.1.1 ISO Certifications

The International Organization for Standardization (ISO) sets a series of worldwide specialized standardization systems. Along with the International Electrotechnical Commission (IEC) they devised standards specifically relating information security management systems (ISMS) known colloquially as ISO 27001.[1]

> [27001] This International Standard has been prepared to provide requirements for establishing, implementing, maintaining and continually improving an information security management system. *The adoption of an information security management system is a strategic decision for an organization.* The establishment and implementation of an organization's information security management system is influenced by the organization's needs and objectives, security requirements, the organizational processes used and the size and structure of the organization.[2]

---

[1]ISO/IEC 27001:2013 Information Technology—Security techniques—Information Security Management System—Requirement.
[2]Introduction, ISO 27001:2013. (Emphasis added.)

© Springer International Publishing AG 2017
K. Williams et al., *The Legal Technology Guidebook*,
DOI 10.1007/978-3-319-54523-3_8

Perhaps the most important thing to understand about ISO 27001, is that, as noted, it is meant to reflect a sort of internal assessment and dialogue between the organization seeking certification and the requirements. So, for example, we might all agree that an earthquake is a catastrophic event, but if your firm or business isn't located in an area prone to earthquakes, you might appropriately determine earthquakes are not a sufficient ISMS risk to merit mitigation. Or, if the business value of certain data is $100,000, a mitigation strategy that requires a $500,000 investment, might not be appropriate.

Independent certification bodies, rather than ISO itself, perform certification audits for ISO 27001. Many of these certifying bodies may also consult organizations to help them prepare for the audit. There are global, regional, and local certification bodies that can suit virtually any organization's size and complexity.

8.1.2 PMP

The key Project Management Professional (PMP) certification standards are articulated in the Project Management Body of Knowledge (PMBOK® Guide) which is promulgated by the Project Management Institute.

In addition to developing a base of substantive knowledge regarding project management, PMP certification requires upwards of 4500 h of actively "leading and directing projects."[3] The Project Management Institute (PMI) oversees the standards, and administers certification exams. However, many universities, and other training organizations, offer preparatory courses for the certification exam.

8.1.3 Six Sigma

Legal Technology

Various Six Sigma tools have been deployed, or referenced, throughout this *Legal Technology Guidebook*. It is a process using quality control tools and managerial strategies to reduce costs and improve efficiencies and quality. A Six Sigma program is based on a key principle of *continuous improvement* through identifying and remediating root causes of defects and variation. Specifically, Six Sigma leverages metrics and statistics to gain insights into the effectiveness and/or impact of processes and process improvements.

For legal professionals, as with myriad other professions, Six Sigma can be used to improve quality and increase customer satisfaction, through: (i) identifying non-value added process steps for cost containment; (ii) understanding the importance of gaining stakeholder and leadership support for your projects; and, (iii) integrating process improvement and quality control for improved results.

---

[3]Project Management Institute, www.pmi.org.

There are four recognized levels of Six Sigma certification, also known as "belts" (yes, like martial artists). From least to most advanced, they are: Green Belt, Yellow Belt, Black Belt, and Master Black Belt.

You may see some organizations that also offer "White Belt" training, this is typically an introduction to Six Sigma precepts, but is not a recognized, or standardized, certification.

As with, ISO 27001, there is not a centralized Six Sigma certification body. Although, the more advanced Six Sigma Belts, as with a PMP certification, require logging time conducting Six Sigma based projects. Readers are encouraged to consider university programs for Six Sigma instruction, or those offered by established organizations, such as the American Society for Quality.

Part III
Technology in Litigation

Chapter 9
Introduction to eDiscovery

At the end of discovery, when all the data is neatly culled, processed and searched, the lawyer must feel like a dog that chases cars. What would the dog do if he caught one? Like the dog, the lawyer has to wonder what will I do with all this stuff? The answer is, of course, get it into evidence in support of a proposition of fact that the lawyer is obliged to prove. To do that, the lawyer has to understand the rules of evidence that will permit or prohibit the consideration of the evidence she will offer.

Unfortunately, the best school to learn the rules of evidence has now closed. Since we, as a country, try less than 1% of the cases filed, learning the rules in the school of hard knocks by actually trying a case has disappeared. Gone are the days when the young lawyer could, by trying a petit larceny, shoplifting case or a "slip and fall" at the grocery, learn from an admittedly tough teacher—the judge—what she was supposed to do and what she could not do when she offered an exhibit or questioned a witness. Of course, there are books available but one supposes there are books on how to roller skate. There is, however, no substitute for getting up on the skates. In the meanwhile, the competent lawyer is going to have to start with the books where the rules of evidence are found.

In the federal courts, the lawyer is lucky-there are the Federal Rules of Evidence that cover every imaginable evidentiary problem. Many states have followed the federal lead and have adopted rules of evidence that are identical to the Federal Rules of Evidence. But, there is a hitch. These rules grew out of centuries of the courts'confronting physical evidence (the gun in a homicide case), the testimony of witnesses, and pieces of paper. The rules therefore speak nearly entirely to this kind of evidence. But, 98% of the communications on the earth are digital and there is no turning back. Thus, with a few minor exceptions, the rules of evidence do not speak to 98% of the evidence the lawyer and the judge confront. Since few of the rules speak specifically to digital evidence, the rest of them must be construed and applied although nearly all these rules that speak to everything that may be admitted, other than digital evidence.

© Springer International Publishing AG 2017
K. Williams et al., *The Legal Technology Guidebook*,
DOI 10.1007/978-3-319-54523-3_9

This pouring of new wine into very old bottles has yielded two schools of thought. One is represented by George Paul and his book, <u>Foundations of Digital Evidence</u>. Paul argues that the rules of evidence were premised on a philosophy of empiricism and the rules that it generated have nothing to do with how the modern world assesses the accuracy of its communications. Paul therefore argues in favor of a radically different approach to the admission of digital evidence.[1]

The second school is "if it ain't broke, don't fix it." Members of this school (including several judges) insist that the old rules will work very well with the new technology, as they have worked with information generated by telegraph messages and copy machines.

While the battle lines have formed, there is stalemate. There is no perceptible movement towards the wholesale revision of the Federal Rules of Evidence to deal with digital information. Thus, like it or not, the competent lawyer will have to deal with the Rules as they are, no matter how poor the fit between the pertinent Rule and the information being offered.

The resulting analysis will often have the lawyer thinking that the courts are constantly asking the wrong question. Her technological competence will convince her that, for example, the metadata in an email will show to scientific certainty that it is from a certain person who was using a certain computer on a certain network. Yet, the question still posed is whether there is independent probative evidence that it is what it is purported to be. But, if we know, as we should, that the metadata establishes to a certainty who created it and where, why doesn't the metadata and our understanding of it require nothing more? The answer is that we do not know, as we should, about how the metadata can prove the email is what it purports to be. Thus, we search for evidence when the metadata, which is staring us in the face, would convince a forensic scientist that the email is what it purports to be and that is was sent by a certain person to a certain person and that the latter opened it. These facts are indubitable and the need for anything further to prove authenticity is wasteful and uneducated. The only real question is how likely is it that the metadata we can see has been manipulated and we should be arguing about that. If the answer is that it is unlikely to the point of impossibility, then the evidence should be admitted without further inquiry. The only thing preventing this from happening is that none of the players dealing with the receipt of the evidence knows enough about metadata and forensic evaluation of digital evidence to understand its significance.

An excellent example of the problem is *United States v. Washington*, 498 F.3d 225 (4th Cir. 2007) which dealt with the admissibility of a report generated by a machine that tested a driver's blood for the presence of alcohol or drugs. The rather silly argument was made that it was "hearsay", defined by the Federal Rules of Evidence 801(c) as the statement of a declarant, not made while testifying, that is offered to prove the truth of its contents. The court quickly rejected the contention that the report was hearsay for the obvious reason that a machine is not a

---

[1] I should note that I wrote the foreword to Paul's book.

"declarant"; a human being can only be a declarant. But, as the dissenting judge pointed out, the report had to be seen as the "statement" of the technicians who performed the test and the defendant should have been permitted to cross-examine them. Id. at 234. It would follow that the cross examination would focus on the real questions presented: did the machine function properly, was is calibrated, and are its result therefore trustworthy enough to be permitted into evidence? Focusing on the hearsay rule obscured the real question presented. We prohibit hearsay because we insist that those who would give testimony be subject to cross examination. That principle has nothing to do with whether a machine produces an accurate result.

This missing of the target when it comes to scientific evidence haunts the questions presented by the offering of digital evidence. For example, imagine Uncle Harry who left two wills, one of which left everything to his son and a second, which left everything to his second wife. The fundamental principle, is that there must be sufficient evidence to support a finding that the document is what it purports to be. Federal Rule of Evidence 901. If there is, the document meets the requirement called "authentication" and may be considered by the finder of fact. In a paper universe that requirement is meet by a testimony of a witness with knowledge, i.e. the attesting witnesses who saw Uncle Harry sign the first will or a notary public before whom Uncle Harry signed the will. Federal Rule of Evidence 901(a)(1). Uncle Harry's long time secretary could testify that she was familiar with Uncle Harry's signature and that the signature is Uncle Harry's. A witness who qualifies as an expert under Federal Rule of Evidence 702 could testify that she has sufficient training and education to opine that there is a scientific basis for comparing signatures, that she used it, and the signature on the will compares favorably with known specimens of Uncle Harry's will.

All of this is well and good but what is Uncle Harry's lawyer sent him a PDF of the second will and Uncle Harry "signed" it using DocuSign or a similar program that permits its users to affix their electronic signatures to electronic documents? Or, even better, supposed he used a website that, armed with artificial intelligence created a will from his responses to his questions which he then signed electronically? What do the rules pertaining to a written will have to do with the will created or signed electronically?

As explained above, one school of thought insists that the present Rules can be made to fit this new reality. But, another school insists that a digital universe mocks the notions of authenticity and hearsay in a paper world. They point out that contracts or sales are made by clicking a box on Amazon and that electronically stored information is produced by machines with decreasing human input. Reliability and trustworthiness are a function of the correctness of the coding of a software program but that correctness is never questioned or tested under the existing evidentiary rules. Shouldn't we be looking at hash values, metadata, and computer forensics which could provide scientific certainty that the electronically stored information—Uncle Harry's digital will—was prepared on a lawyer's computer, transmitted to Harry, where it was opened my Uncle Harry, who signed it using a software program to sign it?

   While this debate goes on, the competent lawyer is going to have to understand the existing rules and how the courts have used them to rule on the admissibility of digital evidence. To that we now turn. A word of warning—the courts differ widely in their interpretations and their opinions are a function of the technology when the judges wrote. Note the dates of the decisions; they may speak volumes of the continued validity of the courts' reasoning. More importantly, there may be emerging a generation of technologically competent lawyers who will take up the struggle of focusing the courts' attention on the scientific basis for the admission of digital evidence. There is no substitute for having the wisdom to identify the crucial documents and the spending the time and money to establish forensically and therefore beyond all doubt that they are what they seem to be and are trustworthy. This does not mean that the lawyer has to hire a forensic expert to get every email in evidence. But, if there is a crucial email, and a question of its authenticity arises, the competent lawyer is well advised to attack that question scientifically, equipped with a deep knowledge of how computers operate and store information. Forearmed the lawyer can gather and present the forensic evidence needed to prove its authenticity. While that is demanding, it makes more sense that arguing about whether a Xerox machine is a declarant.

Chapter 10
Authenticity

1. **The fundamental rule**. A document is not relevant unless it is what it purports to be and, as explained above, a party must therefore produce sufficient evidence to support a finding that it is what it purports to be. Federal Rule of Evidence 901(a).

2. **The role of the judge**. Under Federal Rule of Evidence 104(a) preliminary questions concerning the admissibility of evidence are determined by the judge who is not bound by the rules of evidence, except those with respect to privilege. That rule is in turn subject to Federal Rule of Evidence 104(e) that indicates that the judge's preliminary ruling under Federal Rule of Evidence 104(a) does not limit a party's right to introduce to the jury that is relevant to the weight or credibility of other evidence.

   What this all means is that the judge may consider inadmissible evidence, such as a declaration from a witness, which would otherwise be inadmissible as hearsay, in determining whether Uncle Harry's digital will is authentic. But, there is nothing preventing the judge from concluding that both the digital will and the paper will are both authentic and from counsel for the parties arguing why their client's will is the real will and the other side's will is a forgery. The preliminary ruling that the wills are authentic is just that—a preliminary ruling. Once both wills are admitted, the jury will decide which one should be considered Uncle Harry's will.

3. **Self-authenticating documents**. Fortunately for lawyers, Federal Rule of Evidence 902 takes a group of objects and deems self-authenticating, so that no extrinsic evidence of authenticity is necessary in order for them to be admitted. There is nothing in the Rules that could possibly be construed to limit admissibility to pieces of paper. It would follow that digital versions of these objects should be equally admissible.

© Springer International Publishing AG 2017
K. Williams et al., *The Legal Technology Guidebook*,
DOI 10.1007/978-3-319-54523-3_10

The objects are:
Public documents under seal
Birth and death certificates
Official publications
Newspapers and magazines
Trade inscriptions
Commercial Paper
Acknowledged Documents
Certified copies of public documents.

4. **The Authentication Rules**. If the digital information is not self-authenticating, then the lawyer must resort to the sections of Federal Rule of Evidence 901 that apply to digital information. The three sections most frequently used by the lawyers and the courts are:

901(b)(1) <u>Testimony of a Witness with Knowledge</u>: Testimony [by a human being] that an item is what it is claimed to be.
901(b)(4) <u>Distinctive Characteristics and the Like</u>. The appearance, contents, substance, internal patterns or other distinctive characteristics of the item, taken together with all the circumstances.
901(b)(9) <u>Evidence about a Process or System</u>. Evidence describing a process or system and showing that it produces an accurate result.
We now turn to the cases in which the courts, invoking these subsections of the rules, have ruled upon the most common kinds of digital information.[1]

10.1 The Problems and Their Solutions

10.1.1 Tender of Electronic Information to Establish That the Information was on a Website

Sensitive to the reality that websites can be hacked and altered, many courts to date have been demanding and rejected the notion that printouts from a website are *ipso facto* authentic. *United States v. Jackson*, 208 F.3d 633, 638 (7th Cir. 2000)(tender of web postings rejected because proponent failed to show that they were actually posted by users of the websites rather than herself; proponent's skill as computer user, capable of manufacturing phony websites, noted); *Wady v. Provident Life and Accident Ins. Co. of America*, 216 F. Supp. 2d 1060 (C.D. Cal. 2002) (*Jackson* followed; downloaded material that proponent claimed were from website rejected as not authenticated; no showing of knowledge of who maintained the website, authored the documents or accuracy of their contents); *Whelan v. Hartford Life &*

---

[1]The central and authoritative text is Judge Paul Grimm's remarkably comprehensive opinion in *Lorraine v. Makel Amer. Ins. Co.*, 241 F.R.D. 534 (D. Md. 2007). <u>See</u> Paul Grimm, *Back to the Future: Lorraine v. Makel Amer. Ins. Co.*, 241 F.R.D. 534 (D. Md. 2007).

Accident Ins. Co., 2007 WL 1891175 (C.D. Ca. June 28, 2007) (proponent tendered printouts from Nexis with certification by counsel that print outs were true and correct copies of results of internet search; tender rejected as not authentic even though they bore date stamp and URL; court insisted upon a declaration by personally conducted the search or by company starting that printouts are true and correct copies of information from its website); *Bowers v. Rectors and Visitors of the Univ. of Virginia*, 2007 WL 2963818, at * 1(W.D.Va. Oct.9, 2007) (counsel's representation in affidavit that information was from published websites constituted "abject failure on her part to understand and appreciate a number of evidence rules" including Federal Rule of Evidence 901; effort by counsel said to be reckless and failure to exercise legal judgment).[2]

While some courts in nearly identical circumstances have been more forgiving,[3] counsel who walks into court with a handful of Xeroxed pages from websites excepting them to be admitted into evidence may walk out of the courtroom with them; the jury may never see them. It is, of course, possible that the court will permit counsel to access the website in the courtroom but there is no guarantee that, without more, such as a declaration of authenticity, from the website's owner or operation, that the judge will permit the jury to see what is on the website.[4]

10.1.2   The Internet Archive—"The Way Back Machine"

The Internet Archive, fondly called by fans of Rocky, Bullwinkle and Mr. Peabody, the "Way Back Machine",[5] is an Internet service that saves Internet pages and claims to have 480 billion pages saved to date. It permits a user to actually reconstruct how a webpage looked on a given day. The courts have insisted that the postings on the Internet Archives be authenticated by an affidavit from a representative of the Internet Archive that it retrieved copies of the websites' contents as

---

[2]For an extreme view, see *St. Clair v. Johnny's Oyster & Shrimp*, 76 F. Supp. 2d 773, 775 (S.D. Tx. 1999) ("[A]ny evidence procured off the Internet is adequate for almost nothing").

[3]*Perfect 10 Inc. v. Cybernet Ventures, Inc.*, 213 F. Supp. 2d 1146, 1153–54 (C.C. Cal 2002) (declaration that electronically stored information ("esi") was downloaded from webpages and contained dates on which they were printed met burden of Federal Rule of Evidence 901 because they permitted juror to believe that they are what proponents says they are); *Hood v. Dryvit Sys., Inc.*, 2005 WL 3005612 at * 2 (N.D. Ill, Nov. 8, 2005) (affidavit of counsel that he retrieved documents from website and that addresses stamped on bottom of each exhibit were addresses he retrieved from web site in support of motion for summary judgment was sufficient to permit consideration of documents).

[4]See Ira M. Robbins, *Writings on the Wall: The Need for an Authorship-Centric Approach to the Authentication of Social-Networking Evidence*, 13 Minn. J. Of Law, Science & Technology 1 (2012).

[5]In a cartoon on the Rocky and Bullwinkle show, Mr. Peabody, a very intelligent dog, would use the Way Back Machine to permit him and his owner, a boy named Sherman, to visit history as it occurred whether it was the assassination of Caesar or the outbreak of the Civil War.

they appeared on the dates in question. Compare *St. Luke's Cataract & Laser Inst. P.A. v. Sanderson*, 2006 WL 130242 (M.D. Fl. May 12, 2006) (affidavit of fact witnesses who provided their opinions as to how the Internet Archives works which were in any event incorrect combined with two year old affidavit from Molly Davis, who works for Internet Archives were insufficient to authenticate print outs proponent offered) with *Telewizja Polska USA, Inc. v. Echostar Satellite Corp.*, 2004 WL 2367740 * 6 (N.D. Ill) affidavit of same woman, Molly Davis, attesting that documents were from the Internet Archive and represented contents of website on day in question suffice to establish authenticity).

Note, however, that Molly Davis's declaration, only attests that the website was on the Internet on a given day. Whether or not it was actually posted by the person proponent claimed posted it remains an open question. Thus, as the court pointed out in *Novak v. Tucows, Inc.*, 2007 WL 922306 * 5 (E.D.N.Y. Mar. 26, 2007), while a proponent may assert that the website was on the Internet on a given day, Ms. Davis still lacks personal knowledge of who posted it and "information posted on the Way back Machine is only as valid as the third party donating the page decides to make it—the authorized owners and managers of the archived websites play no role in ensuring that the material posted in the Way back machine accurately represents what was posted on their official websites at the relevant time." Failure to produce sworn statements by employees hosting the sites from which plaintiff printed the pages doomed the argument that the print outs from the "Way back machine" were authentic.

10.1.3 Communications Using the Internet

Emails can be authenticated under Federal Rule of Evidence 901(b)(1) by testimony of a witness with knowledge that the item is what it is claimed to be. Therefore, an affidavit from the recipient may suffice to establish the authenticity of the email that was sent. *Fence v. Feld*, 301 F. Supp. 2d 781, 809 (N.D. Ill. 2003); *Maier v. Pac. Heritage Homes, Inc.*, 72 F. Supp. 2d 1184,1190 (D. Or. 1999) (affidavit from recipient suffices to authenticate authenticity of what was sent).

Similarly, testimony as to the exchange of emails or online chats by one of the parties who exchanged them suffices to authenticate them. *United States v. Gagliardi, 506 F.3d 140 (2d Cir. 2007)*.

But, what if neither sender nor recipient is available? This can happen in a criminal case where the defendant cannot be compelled to testify and the other party to the email is unavailable. This happened in *United States v. Safavian*, 435 F. Supp. 2d 36 (D.D.C. 2006) where the author of the emails to the defendant, a lobbyist named Abramoff, had a Fifth Amendment privilege not to testify and the defendant, of course, could not be compelled to testify that he sent or received emails form Abramoff. The government subpoenaed Abramoff's email to and from the defendant from the law firm where Abramoff worked and offered them into evidence. The court concluded that the emails could be authenticated by their distinctive characteristics or circumstantially because, for example, the author of one of the disputed emails knew information that would have been known to the person who appears on the email to be the sender of the recipient.

As the court noted, Federal Rule of Evidence 901(b)(4) permits authentication by "appearance, contents, substance, internal patterns, or other distinctive characteristics, taken in conjunction with the circumstances". The emails had the following distinctive characteristics:

1. email addresses contain the @ symbol, known to be part of an email address, meaning that it was an email;
2. names of person who had that email address;
3. "ambramoffj@gtlaw.com" and recipient, David.Safavian@gsa.gov were names of the sender and recipient in "To" and "From" headings in the emails;
4. "signatures" appeared on the emails, i.e., what were probably JPG images of their signatures;
5. The email spoke of independently identifiable matters such as the employment of the sender and recipient.

These "distinctive characteristics" served to establish that the emails were what they purported to be. Once an email was authenticated in this manner, the jury could compare a questioned email with the authenticated one, satisfying Federal Rule of Evidence 901(b)(3), which states that trier of fact may compare a document, claimed to be authenticated, with one that has been authenticated and determine the authenticity of the former by comparison with the latter. Id.at 40–41. Thus, once the judge concluded that a certain email was authentic because of its distinctive characteristics, the jury could use the authenticated email to determine the authenticity of another email because, for example, it had the same email address for a person that appeared in the authenticated email, Finally, the possibility of alteration did not in itself defeat authenticity, Id.at 41.[6]

A similar result was reached in *United States v. Siddiqui*, 235 F.3d 1318, 1322–23 (11th Cir. 2000). The tendered emails bore Siddiqui's email address and that was the address that the computer yielded when the person to whom Siddiqui sent the email hit the reply function on the correspondent's computer. The context of email sent indicated that it was from someone who would have known information that Siddiqui knew. Emails sent to two men named Yamada and Von Guten referred to author as "Mo"; "Mo" was Siddiqui's nickname. Yamada and Von Guten testified that, when they spoke to Siddiqui soon after their receipt of the email, Siddiqui made the same request as had been made in the email. All of this served to permit the conclusion that emails were what they purported to be.

The cynical would note that "distinctive characteristics" can be spoofed, i.e. copied and then use in a forged email. While that is certainly true, the possibility of alteration cannot serve in itself to require finding all digital evidence inadmissible

---

[6]Note that government witnesses were precluded from testifying that a certain email was to or from one witness; the jury would draw that inference if it saw fit from the contents of the emails.

unless the courts are ready to require forensic evidence, eliminating any possibility of forgery or alteration, as a condition of the receipt of all digital evidence. Whether a compelling case can be made that they should, given the prevalence of fraud, forgery and misattribution on the Internet the courts are not there yet. Presently, distinctive characteristics of the email as specified in *Safavian*, and circumstantial evidence permitting the conclusion that it is what it appears to be, will suffice.[7]

10.1.4 Text Messages and Instantaneous Communications

The authentication of text messages, which may be called "tweets" because transmitted by a Twitter account, is analyzed using the same technique of using distinctive characteristics and circumstantial analysis. *Adams v. Disbennett*, 2008 WL 4615623 (Ohio Ct. App. Oct. 20, 2008) is a good example.

It has been said that love is grand and divorce is about two grand. Just like Frankie and Johnny, Adams and Disbennett used to be lovers. Adams loaned Disbennett money and, when they broke up, Adams, sadder but wiser, wanted it back. He had several messages that the two exchanged and offered them to prove that Disbennett had acknowledged the debt. He testified that "Adams man 003" was his screen name and that Disbennett's real name appeared on his computer when he typed in her screen name (pardon my French) "sexy bitch 43302." Adams had copies of the printouts of the messages, which were both tweets and emails and he pointed out that in them there was information he (perhaps euphemistically) called "private".

Despite Disbennett's objections, the court admitted the emails and messages finding first that authentication could be proved by a person with knowledge and Adams, of course, knew Disbennett's screen name. It then followed that, since authenticity can be proven circumstantially, it was highly probative, that, when, for example, a person gets a text, and uses the reply function, that function populated the sender box with the words "sexybitch 43302" which Adams knew to be Disbennett's screen name. An email to Adams referred to him by a particular name and, as Adams testified, the sender made the same request of him in the email that she had made in person. All of this convinced the court that there was sufficient evidence that the texts and emails were what they purported. Again, the combination of distinctive characteristics and circumstantial evidence permitted the conclusion that there was sufficient evidence of authenticity.[8]

---

[7]Note that in that context, the burden of establishing authenticity is said to be a light one. *United States v. Appolon*, 715 F.3d 362, 371 (1st Cir. 2013) (relatively undemanding requirement of proving authenticity is met by proof showing reasonable probability evidence is what it claims to be).

[8]Note the similarity of this conclusion to the conclusion of authenticity in criminal cases where the defendant uses the computer to commit the crime, by, for example, downloading child pornography and then accepting the invitation of what he thinks to be a fellow predator to have sex with a

10.1.5   Social Media Postings and Tweets

It is becoming common to use social media postings and many cases are being lost or won on the basis of their contents. Again, in the absence of forensic evidence the proof has had to be circumstantial. For example, in *Maryland v. Griffin*, 19 A.3d 415 (Md. 2011), a page from the defendant's girlfriend "My Space" page[9] was offered into evidence. The defendant's nickname was "Boozy" and the page contained a threat: "Free Boozy. Just remember snitches get stitches." It was offered by the prosecution to explain why a government witness who had in the first trial declined to provide incriminatory evidence offered it in a second trial. The only witness offered to authenticate the page was a detective who saw the screen, bearing this page, in a computer in the girlfriend's apartment, and also saw references on the page to her age, date of birth, and a picture of her and Boozy embracing. Even though the girlfriend testified, she was not asked about the My Space page.

The court held that, in light of the ease with which someone else could have created it, the evidence was insufficient to authenticate the page and its admission was error. The court suggested other means of authenticating such a page, including, of course, forensic evidence. Id. at 428.

The same court confronted the issue of authentication in *Sublet v. Maryland*, 113 A.2d 695 (Md. 2015) that involved social media postings and text messages in several distinct cases.

(1)   The court held that, when a witness testified that she permitted others to know her password, and had the experience that other persons "hacked" into her site and wrote things she did not authorize, and that she did not write what appeared on a final page of a series of Facebook entries, the trial court did not err in refusing to admit the final page. When this witness denied having personal knowledge of the creation of item to be authenticated, its authenticity was undermined. The last page lacked any distinct characteristics. In fact, the posts on the last page were disconnected entirely from the posts on the first page which the witness admitted authoring. Additionally, the posts on the last page did not refer to posts on the earlier pages.

(2)   After a forensic expert testified that he had used special software to retrieve message from an iPhone and determined that two persons who had chat names "They LovingTC" and "OMGitsLoco" were the participants in a series of tweets. The expert produced an analysis of the contents and times of the Tweets. Photos on one account were photographs of a man named Harris whose Twitter name was "They Loving TC". The contents of the messages

---

(Footnote 8 continued)

child. The fellow predator is actually a police offer who, by email, invites the defendant to meet him at a certain place. When the defendant arrives there, it permits the conclusion that the emails between the two of them are authentic.

[9]"My Space" was an antecedent of Facebook.

indicated that the participants knew of and were involved in the assault, which involved only a limited group of people. The messages occurred on the day the assault took place and one of the participants, in concocting the plan to assault another person used the name "They Loving TC". (One direct message from "They Loving TC:" "Shit going to get real tomorrow."). The jury could therefore find that Harris was "They Loving TC" and the direct messages between him and "OMGitsloco" were authentic. Once the judge so concluded, he could also conclude that Tweets by "They Loving TC" were by Harris, since they were in the same time frame. Indeed, the Tweets were within minutes of the Tweets between "TheyLoving TC" and "OMGitsLoco." "Based on the temporal proximity of the directs messages and the Tweets that were already authenticated, a juror could find that the Tweets were authentic."

(3) The victim of the assault testified that while she was recovering from being stabbed she received Tweets, and attested that they were from defendant (whom she had dated for a year). The court held that they were authenticated when (1) they were sent at a time when few people knew of the stabbing and referred to it: (2) expressed the author's remorse for the stabbing; (3) all the messages were in Spanish and the victim and the defendant were Hispanic; (4) defendant called her by phone thereafter, and she found a note from him in her apartment in which he also sought her forgiveness.

A similar interconnected string of facts was held to authenticate a My Space page in *Tienda v. Texas*, 358 S.W.3d 633 (Ct. Crim. App. Tx. 2012). In this murder case, a My Space page was admitted into evidence to show that the defendant was the murderer. There were photos on the page and instant messages and comments linked to the account. There were also two links to music. Specifically, there were:

1. Two page profiles were created by Ron Mr. T and the third by "Smiley Face" which is defendant's nickname;
2. The account holder lived in "D Town," i.e. Dallas;
3. The account gave his names ronnietiendair @ _____; and "smiley's shit." Tienda was the defendant's last name and, as just noted, his nickname was Smiley;
4. Pictures on site resembled him;
5. One of the music links was to the music played at the murder victim's funeral;
6. The person depicted on the site had gang affiliate tattoos and was making gang related gestures;
7. The holder of the account said in a posting "I kill to stay rich;"
8. Messages between account holder and others included specific references to other persons who were known to be present during the shooting;
9. The account holder threatened anyone who ratted on him;
10. The author of posts complained about having been in lock down and now had to wear an electronic monitor.

The court held that the admission of the My Space page was correct because the following facts provided "ample evidence" to support a finding that pages were authenticated and created by the defendant:

(1) the photographs of him with his unique arm, body, and neck tattoo and his distinctive eyeglasses and earrings;

(2) the reference to the victim's'death and the link to the music played at the victim's funeral;

(3) the references in the My Space page to his gang ("Tattoo Blast") and the distinct gang gestures in the photographs;

(4) the messages referring to the shooting and to a certain person (a government witness at the murder trial) as a snitch;

(5) that the defendant was on electronic monitoring as the messages indicated, along with photo of him showing off the monitor on his ankle;

(6) the identifying references on the page to "Ron Mr. T" or "Mr. Smiley Face." whose email address (ronnietiendaajr@—) was defendant's.

The court distinguished the *Griffin* case on the obvious grounds that the wealth of circumstantial evidence in the case before it surely permitted a reasonable person to conclude that the My Space page was the defendant's. See also, *United States v. Vayner*, 769 F.3d 125, 131 (2d Cir. 2014). When government tried to establish that a certain email address was defendant's, and produced a page from what it called the Russian equivalent of Facebook, indicating that the person identified on the Facebook posting had an email address that was the email address used by the perpetrator of the crime its proof of authenticity failed for lack of showing that the defendant authored the posting).

The extraordinary length to which the courts in these cases went to use distinctive characteristics and circumstantial evidence to establish authenticity are testaments to their hard work and scholarship. But, one must ask why it never occurred to the prosecutors to get search warrants for the computers and then have a forensic expert examine them and speak authoritatively and scientifically as to why the conclusion that the pages were created by a person who then posted the information on them. Had they done that, what took these courts so much time and effort to evaluate could have been so easily avoided. The technologically competent prosecutor will have to learn enough about the available technology to understand how computers operate and how their use leaves information upon them that permits courts to conclude that that the information is what the prosecutor claims it to be—Tweets from a certain person or that person's Facebook page. Surely, the obligation to establish the defendant's guilt beyond a reasonable doubt and the necessity to prove a case efficiently and intelligently requires at least that.

10.1.6 Computer Generated Data

Databases now contain much of the electronically stored information that the world generates and the courts have therefore had to confront the authenticity of information, usually business records, that are generated from a computerized database. The pertinent Federal Rule of Evidence is 901(9) that indicates that evidence describing a process or system and showing that it produces an accurate result will authenticate the product of the system or process.

As discussed above, the dissent in the *Washington* case correctly pointed the production of information from a software process is an implicit representation that the process was correctly coded and designed to achieve the result that is now being offered into evidence. The Rule, just quoted, requires no less but the courts have been quite forgiving in their admission of computer-generated information. In *Linderoth Associates v. Amberwood Development, Inc.* 2007 WL 269851 (D. Az, Sept 2, 2007), for example, the court noted that the custodian of the record need not have technical knowledge of how the computer functions nor be the person who created the file. It suffices that the witness understood how the software recorded the start and modification dates for a certain file, had reviewed it, ascertained its creation date, and its last modified or completed date. This was enough to authenticate the file. In one instance, however the court took a radically different and demanding approach. In *In re Vee Vinhee*, 336 B.R. 437 (US Bank. App. Panel 9th Cir. Dec. 16, 2005). American Express only offered into evidence a printout of an American Express computerized record that indicated the bankrupt's debts but the witness testified only that the print out was marked "duplicate copy." An American Express witness testified that knew nothing about how the software created the record. The court found this deficiency fatal and declined to admit the record. The court held that there was no proof of the witness's qualifications to testify nor how American Express conducts its business in reliance on the software and its capacity to create, retain, and retrieve the information that the print-out reflected.

The court concluded that a mere identification of the software used would not work. Instead, American Express should have shown how access to the database was controlled, how changes in the database were logged or recorded, as well as the structure and implementation of backup systems and security procedures for insuring the integrity of the database. The court noted that the ease with which e data can be changed in our world increases the importance of understanding the audit procedures used to insure the integrity of the data.

The court also indicated that, as explained above, Federal Rule of Evidence 901 (b)(9) permits evidence that describes a process of system and shows that it produces an accurate result suffices to establish authenticity. But, at a minimum, that requirement must be read to require proof of the safeguards to insure accuracy and identify errors. This includes details regarding computer policy and system control procedure, including control of access to the database, recording and logging of changes, backup practices and audit procedures.

Finally, the court held that witness who said: "The computer does not change numbers" was not a qualified witness. He had no knowledge of steps taken to guard against changes in the data in the system American Express used. His representations that noted the make and models of the equipment and software used to generate the data, that both were updated periodically and were standard for the industry were insufficient to authenticate the print out from the database.

It remains to be seen whether this case will gather adherents to a more realistic and scientifically based approach to authenticity. In the meanwhile, the technologically competent lawyer will remember all the blunders counsel for American Express made and never take for granted that proffering the output of a database without more will render it admissible as authentic. Counsel must be prepared to establish from a capable and competent witness (1) how the database works, (2) how is information input, (3) how the software is programmed, (4) why it is certain to achieve the required output, (5) how, if it all, are the results audited, and (6) what steps have been taken to keep it secure from the moment it is created to the moment it is offered. If technologically competent counsel is not ready to do that, the result may be as surprising as the result in *In re Vee Vinhee* was to the lawyers for American Express.

10.1.7 Business Records

Federal Rule of Evidence 902 (11) mirrors Federal Rule of Evidence 803(6)(A)–(C). A record will be authenticated and admitted if a custodian or "other qualified" person meets the requirements of Federal Rule of Evidence 803(6)(A)–(C) Those requirements are:

(1) that record was made at or near the time by—or with information transmitted by—someone with knowledge;
(2) the record was kept in the course of a regularly conducted activity of a business;
(3) making the record was a regular practice of that activity.
(4) neither the source of information or the method or circumstances of preparation indicate a lack of trustworthiness.

All of these conditions may be shown by the testimony of the custodian of the record or another qualified person or by a certification that complies with Federal Rule of Evidence 902(11). Federal Rule of Evidence 803 (6) (D). But, to be a custodian or other qualified person who can make this certification, the person must have sufficient knowledge of the record-keeping system and the creation of the contested record to establish its trustworthiness; Thus, a certification that the custodian collected the records from a particular source is insufficient. *Rambos v. Infineon Tech. AG*, 348 F. Supp. 2d 698, 703 (E.D. Va. 2008) (proponent's affidavits insufficient because "[they} fail to make any reference to the declarant's knowledge, or even awareness, of the record-keeping practices of the company that produced the documents").

As to the other requirements, to qualify as being "kept in the course of a regularly conducted activity of a business" means that the mere presence of a document ... in the retained filed of a business entity do[es] not it by itself qualify the document as a record of regularly conducted activity." Id. There must be a showing that the record is made pursuant to established procedures for the routine and timely making and preserving of business records, and is relied upon by the business in the performance of its functions. Id.

To meet the requirement that making the record was a regular practice of that entity it is not enough that a particular employee regularly makes and keeps records as part of her regular practice; it must be the regular practice of the business as a part of its regularly conducted activity to make and keep the record at issue. See *Palmer v. Hoffman*, 318 U.S. 109 (1943) (rejecting claim that certain investigative report, routinely done, was a business record; it is not the regularity with which an entry is made but whether all such entries were routine reflections of the day to day operations of a business).

Thus, email text messages, voice recordings may fail to meet these requirements. That people within an organization communicate with each electronically does not mean that their communications are business records. *Rambos*, 348 F. Supp. 2d at 706 (email chains in which there were emails from persons within and without the business could not qualify as business records unless each participant in the chain was acting in the course of regularly conduced business; *United States v. Ferber*, 966 F. Supp. 90, 98 (D. Mass. 1997). *New York v. Microsoft Corp.*, 2002 WL 649951 * 2 (D.D.C. April 12, 2002) (email recounting meeting might have been kept in course of business but there must also be a showing that it was the regular practice of employees to write and maintain such emails).

There is, of course, no rule that bars electronic communications from being deemed authentic as business records. Salespersons who text their orders in or enter them into a database for processing certainly seem to be creating documents that are created in the regular course of business and putting them into a system that uses those texts or entries in the regularized practice of this business activity. it is the obligation to create the record and its use in the daily operations of the business that are the crucial conditions. As the court explained in *Sea-Land Serv. Inc. v. Lozen Int'l*, 285 F.3d 808 (9th Cir. 2002) when speaking of bills of lading that had been generated by a computer:

> The bills of lading are business records. Rule 803(6) allows the admission of business records when "two foundational facts are proved: (1) the writing is made or transmitted by a person with knowledge at or near the time of the incident recorded, and (2) the record is kept in the course of regularly conducted business activity." *United States v. Miller*, 771 F.2d 1219, 1237 (9th Cir.1985).

> The first element of that test is met here because, although the physical documents were not generated when the parties contracted for the shipment of Lozen's grapes, they were produced from the same electronic information that *was* generated contemporaneously. For the purposes of Rule 803(6), "it is immaterial that the business record is maintained in a computer rather than in company books." *United States v. Catabran*, 836 F.2d 453, 457 (9th Cir.1988) (citation and internal quotation marks omitted). Rule 803(6) allows for the admission of a "data compilation, in any form," so long as the compilation meets the requirements of the rule. *Id.*

> Likewise, the second element of the test is met, because the information on the bills of lading is kept in the course of Sea-Land's regularly conducted business activity. The bills of lading are those that would have been issued in the regular course of Sea-Land's business had Lozen not requested that the transaction be performed electronically.

Sea-Land Serv., Inc. v. Lozen Int'l, LLC, 285 F.3d 808, 820 (9th Cir. 2002).

Rendering the electronically stored information authentic and admissible is therefore part of a two stage process that requires answers to two distinct questions: (1) is the record what it purports to be and (2) was it kept in the ordinary course of business and was it a regular practice of the business to make it and to keep it?

The technologically competent lawyer will appreciate how imparting to the court a technical and scientific understanding of the generation of the putative business record may meet both of these requirements if she explains to the court carefully (1) how the computer system being used generated and stored the record within the depository in which it resides and from which it was taken, (2) why the employees of the organization were required to create and keep the record, and (3) how they were used in the daily operation of the business.

10.1.8 Self-authentication

Federal Rule of Evidence 902(4) provides that government records ("copies of official records) are self-authenticating.). Additionally, Federal Rule of Evidence 902(5) indicates that official publication purporting to be issued by a public authority are also self-authenticating. As the government converts its massive informational and other database to a digital format, counsel can offer into evidence a remarkable amount of information that will quickly and easily be authenticated and admitted.[10]

Conclusion

Hunting for how to authenticate digital equipment seems to this point in the case law to be like cutting down a tree with a penknife when there is a chain saw nearby. We lawyers and judges have somehow created the most difficult, circuitous way of answering a ridiculously simple question. is this thing what is it is supposed to be? The technologically competent lawyer can quickly pick up the chain saw and cut to the heart of the matter: here is the forensic, scientific evidence that shows that this electronically stored information is what we say it is. The competent lawyer can answer that question quickly and indubitably if she will only take the time to learn why the scientific answer can be produced quickly and correctly. In the meanwhile, we seemed to be doomed to cutting down trees with pen knifes, afraid to turn on the chain saws. The only good thing we can say about that process is that it will preserve the forest, although it will waste a fortune in the client's money and the court's time.

---

[10]Note that public records are also an exception to the rule against hearsay. Federal Rule of Evidence 803(8).

Chapter 11
The "Ethics" or Lack of Them During Negotiations

One of the most significant of the many achievements of the Sedona Conference was the issuance of its Proclamation on Cooperation. It is startling to realize that there was opposition to the Proclamation's central tenet—that lawyers do not violate any ethic proscription when they are transparent and cooperative with opposing counsel—on the grounds that it was inconsistent with the lawyers' obligation, discussed above, to provide competent representation. The opposition came from those who engraft onto that rule words that do not appear in it—the lawyer must be a zealous advocate and consider the supposed obligation to be zealous to require the lawyer to give not one jot or tittle in discovery that the rules require. This kind of lawyer would never even consider discussing with opposing counsel which custodians are most likely to have the most relevant information, what the format or production by both sides should be, and how their clients' information systems actually operate. This attitude makes the adversary system an absolute in itself and dictates that no actions should be taken by the lawyer that lessens the burden of the opposing party, no matter how much money the client can save by a more cooperative attitude. This attitude is, of course, utterly inconsistent with the central motivation that, for example, animated many of the 2006 and 2015 amendments to the Federal Rules of Civil Procedure—that the meet and confer required by Federal Rule of Civil Procedure 26(f) should be a substantive and meaningful discussion of the all the problems eDiscovery will create as to, for example, format of production, or handling privilege claims. While the Federal Rules do not mandate cooperation, any lawyer worthy of the name knows that a meaningful and useful 26(f) conference cannot possibly take place unless both sides are committed to at least a minimal level of cooperation and transparency.

Unfortunately for those who believe that there is no inconsistency between cooperation and competent representation, the rules pertaining to negotiations are no help at all. They are reflective of the adversary system being deemed an absolute in itself that defines the honest behavior of a lawyer as being little more than avoiding reprehensible behavior when dealing with opposing counsel or, as the Rules put it, any third person.

© Springer International Publishing AG 2017
K. Williams et al., *The Legal Technology Guidebook*,
DOI 10.1007/978-3-319-54523-3_11

More specifically, of Model Rule of Professional Conduct 4.1 only prohibits the lawyer from making a false statement of fact or law. Her obligation to disclose a material fact only arises when it is necessary to avoid assisting a criminal or fraudulent act. Even then, that obligation yields to the responsibility imposed by Model Rule of Professional Conduct 1.6 to protect the confidences and secret of a client. Reading the two rules together, a lawyer is obliged to disclose a material fact only when it is necessary to: (1) prevent reasonably certain death or substantial harm; (2) to prevent the client from committing a crime of fraud that is reasonably certain to result in substantial injury to the financial interest of another, and in furtherance of which the client used the lawyer's services, and (3) to prevent, mitigate, or rectify substantial injury to the financial interests of another that will or has resulted from client's commission of a crime or fraud in furtherance of which the client used the lawyer's services.

Obviously, there is not a word in these rules that could possibly advance an argument that a lawyer has some duty to be honest and transparent during a meet and confer. The only disclosure that she is obliged to make is to prevent reasonably certain death, or substantial bodily harm, or to prevent or rectify a crime of fraud that the client will or has committed, using the lawyer's services. The Comment to Rule 4.1 allows for the possibility that: "Misrepresentations can also occur by partially true but misleading statements or omissions that are the equivalent of affirmative false statements." But, that is cold comfort for those who would insist that lawyers should be transparent and cooperative if the 26(f) process is to work.

The only other obligation imposed on a lawyer to be honest is the requirement under Model Rule of Professional Conduct 8.4 that she avoid engaging in "conduct involving dishonesty, fraud, deceit or misrepresentation." Again, this rule which speaks of avoidance of certain behavior cannot possibly be construed as obliging any kind of openness, candor or transparency during the meet and confer process.

It might be said that a meet and confer is a form of negotiation but that is of no help. There are no rules pertaining to a lawyer's behavior during negotiations other than the ones just discussed. Indeed, the Comment [1] to Model Rule of Professional Conduct 4.1 indicates that it is generally accepted that statements made during negotiations, such as a party's intentions as to an acceptable settlement of a claim, are not deemed "statements of fact." If the meet and confer is deemed to be a negotiation, this Comment could be construed by the cynical to mean that nothing said during it by the lawyers is subject to the requirement of not making a false statement of material fact when they are speaking of what their client will or will not do during the discovery process. Under this theory, the law of the market place, or perhaps, the law of the jungle, prevails. Thus, those looking for some ethical proscription that could serve as a basis for lawyers' cooperating in the 26(f) process will not find it in the pertinent rules of the profession.[1]

---

[1]For a fascinating discussion of whether a lawyer has any obligation to correct an opposing lawyer's mistake during a 26(f) conference see Ethics and Professionalism in the Digital Age, A Symposium of the Mercer Law Review, 60 Mercer L. Rev. 845, 865–893 (2002).

Note, however, the interesting tension between the limited obligations lawyers have under these Rules and several cases that imposed a much greater obligation upon those very same lawyers. See *e.g. State ex. rel. Nebraska v. Addison*, 412 N. W. 2d 88 (Neb. 1987) (failure to disclose liability policy of which opposing counsel was unaware was deceitful); *Virzi v. Grand Trunk Warehouse & Cold Storage Co.*, 571 F. Supp. 507 (E.D. Mich 1983) (failure to disclose plaintiff's death was violation of obligation to be candid with court and counsel); *Kentucky Bar Assoc. v. Geisler*, 938 S.W.2d 578 (Ky. 1997) (*Virzi* followed; ABA Opinion 95–397 that counsel is obliged to advise opposing counsel of death of client followed). Thus, counsel had better to be very careful. Behavior condoned by the ethical rules may be the very behavior a court finds reprehensible. Counsel should remember that courts apply the law, not the Model Rules of Professional Conduct.

Chapter 12
Technology Assisted Review

Lawyers and legal support professionals routinely deploy technology, in one form or another, to conduct document reviews. So, what exactly is meant by the terms Technology Assisted Review (TAR, or predictive coding)?

Linear Review

Let's first, consider the traditional uses of legal technology applied in **linear review**. Linear review involves a review team moving through documents, one by one, in a sequential way. That sequence may be determined by the order in which the documents were collected, by a priority assigned to various custodians in the case, or by any number of other possible criteria. In a linear review, each document will be looked at by lawyer or paralegal at least once.

As TAR is not appropriate, or necessarily desirable, in most discovery contexts, there are, not surprisingly, advanced modes of linear review that are meant to increase both the accuracy and the efficiency of the process. These advanced linear review strategies may be categorized as follows: (i) like document grouping—similar or related documents are assigned in batches to maximize the expertise of given team members, and concentrate learning and related efficiencies; (ii) forced coding—setting rules which automatically force partial or complete coding of certain documents based on decisions made with other documents. For example, if a document attached to an email is deemed responsive, the attaching email may automatically be designated responsive as well; (iii) conceptual review—typically defined as the use of advanced analytics and concept clustering technologies to automatically identify groups of documents that, are largely similar in content and, therefore, likely to be coded the same way.

TAR

TAR and predictive coding involve the use of machine-learning technologies to help make faster, more accurate decisions than typically possible in linear review.

© Springer International Publishing AG 2017
K. Williams et al., *The Legal Technology Guidebook*,
DOI 10.1007/978-3-319-54523-3_12

Particularly, minimizing time wasted in determining that certain documents are *not responsive* to the given review. These technologies, primarily, operate in two ways, which may be applied in tandem in certain products:

1. Automatic Document Prioritization Each review decision is used to prioritize the remaining unreviewed documents based on their expected responsiveness. The documents most likely to be responsive are assigned first. As the review progresses, responsiveness decreases. A determination should be made at the outset, preferably in cooperation with the receiving party, regarding what sampling of the remaining corpus should be conducted once reviewers stop encountering responsive documents. The aim, again, being to minimize any review of not responsive material.

2. Substantive Coding Suggestions Machine-learning may be used to apply decisions at the coding level, with training sets providing the basis to calculate the likelihood of each document receiving a given substantive designation or "tag." This information can be used to not only make predictive assessments, but also to highlight where human reviewers have reached decisions that are unexpected, or incongruous, and might, therefore, warrant additional scrutiny.

Statistics and Metrics

The technologies outlined above are derived from a combination of many individual disciplines, including linguistics, data retrieval, database design, and statistics. While metrics are used to measure particular attributes of a given data set, or population, statistics are numerical characterizations made about a larger population based only on a sample of that population.

We have, thankfully, not yet reached a point where legal professionals must be statisticians in order to ethically discharge their duties, but we are swiftly approaching a place where a degree of comfort with basic terms and concepts ought to be expected. This is particularly the case when lawyers are using TAR.

A key metric in TAR is known as **richness** (or "prevalence") this is the percentage of responsive data existing in the overall review population. In short, the richness of a population will enhance the statistical **significance** of the predictive coding model. In statistics, significance shows us how likely variation is a result of chance. We will look at how increasing richness, increases the relative reliability of our outcomes from a statistical perspective—putting aside the benefits that might also be derived for the linguistic aspects of the given technology being employed.

Most of us are familiar with the concept of a **margin of error**, if not **confidence levels** and **confidence intervals**. But, what do these terms mean in the context of prevailing TAR and predictive coding technologies?

The margin of error (also termed confidence interval) tells us the degree of sampling error associated with a statistical test. As an example, if by randomly sampling X number out of a total population of 1,000,000 documents, we find that 10% of those documents are responsive, then sample size X will determine our margin of error. The

Population 1,000,000	Sample size (X)	Margin of error (Confidence Interval)	Confidence level (%)
	100	9.8	95
		12.9	99
	1000	3.1	95
		4.08	99
	10,000	0.98	95
		1.28	99
	100,000	0.29	95
		0.39	99

Fig. 12.1 .

larger X is, the smaller our margin of error will be. The confidence level then tells us *how* confident we can be in that expectation of 10% prevailing across the whole population (plus or minus the given margin of error).

The table in Fig. 12.1, shows us how these factors interact. By sampling 1000 documents, we can predict the rate of responsiveness across the entire document population, with a margin of error of ±3.1% *and* a confidence level of 95%.[1]

Assuming, as referenced above, we found a 10% rate of responsiveness in the sample that means 6.9–13.1% of the total populations is likely to be responsive (95% likely, to be exact). In other words, anywhere from 69,000 to 131,000 documents might be responsive—a 62,000 document variation. That means we could almost double the number of responsive documents (69,000 × 2) and still be within our prediction. In most circumstances, this would be an unsatisfactory degree of unpredictability.

Of course, one way to address this is to increase the sample size. Sampling 10,000 documents of our 1,000,000 population, we can decrease the margin of error to 0.98%, as shown in Figure x. Now, our model tells us anywhere from 99,000 to 101,000 documents are likely to be responsive. This is degree of variation should be much more acceptable in terms of substantive outcomes as well as being statistically significant, as defined above.

But, another method, which should be pursued in every instance, is to increase the richness of the population. This can be achieved by advanced culling strategies, and in cooperation with the party, or parties, requesting the data. Let's suppose, now that our 1,000,000 document population is the end product of an effective culling strategy, such that our 1000 sample indicates 30% responsiveness. The margin of error is still ±3.1%, which now means we can be 95% confident that

---

[1]NOTE: When you make the confidence level higher, you also increase the margin of error, unless you also increase the sample size.

26.9–33.1% of the total document population is likely to be responsive. It is the same 62,000 document variation, of course, but now the relative magnitude is diminished by 3 times.

Company	Product
kCura	Relativity
Recommind	Axcelerate
Nuix	Nuix
Exterro	Predictive Intelligence
FTI Consulting	Ringtail
Consilio	Backstop

Part IV
Strategies for Achieving Competence

Chapter 13
Transparency

The ABA Model Rules of Conduct Preamble states, "As advocate, a lawyer zealously asserts the client's position under the rules of the adversary system."

As lawyers, zealous advocacy is not only *what* we're taught, but often defines *how* we're taught in the intensely competitive pedagogical approach of most law schools (e.g., Socratic method classroom instruction, and Moot Court competitions.) The prevailing "group project" model in MBA programs is still anathema at most law schools. It is, however, critical for lawyers to develop a capacity to discern when effective advocacy calls for adversarial tactics, and where transparency and/or cooperation are required—routinely, a combination of these strategic approaches should be deployed in the course of one matter.

Practical technological issues inherent to modern legal practice may call for an increased degree of transparency. Adopting this posture is not in conflict with zealous advocacy. To the contrary, transparency is sometimes a critical component in fulfilling one's ethical obligations and, in fact, is often required to most effectively advance the client's interests.

13.1 Defining Transparency

In reality, transparency should operate in many aspects of legal practice, including, billing, administration, and project management. But, what do we mean when we refer to it in the framework of an adversarial process? In the context of direct advocacy, transparency serves as an integral component of cooperation, which should, in turn, serve to minimize costs and effort devoted to the resolution of non-substantive aspects of any dispute or negotiation. Remember that, in fulfilling the role of advocate, lawyers also have an ethical obligation to maintain reasonable

© Springer International Publishing AG 2017
K. Williams et al., *The Legal Technology Guidebook*,
DOI 10.1007/978-3-319-54523-3_13

fees and costs.[1] A key driver in achieving this goal is efficiency. Transparency and cooperation ought to be viewed as vital tools in realizing this goal.

In The Sedona Conference's *Case for Cooperation*, the imperatives created by exponential growth in electronically stored information ("ESI") is explained:

> ... ESI has vastly increased the quantities of available information and the way it can be accessed. With almost all information electronically created and stored, there has been an exponential increase in the amount of information litigants must preserve, search, review, and produce. ESI is often stored in multiple locations, and in forms difficult and expensive to retrieve. *These reasons compel increased transparency, communication, and collaborative discovery. The alternative is that litigation will become too expensive and protracted in a way that denies the parties an opportunity to resolve their disputes on the merits.*[2]

Effective cooperation requires that all parties manifest sufficient transparency, or candor, to minimize costs and maximize results.

13.2 The Growing Need for Transparency in eDiscovery

In attempts to store, manage and analyze vast quantities of data for potentially relevant ESI, litigants may rely on strategies ranging from, searches and filters, to indexing and predictive coding. Often, more than one of these tactics will be employed, and the tools that support these efforts may be widely commercially available, or proprietary and bespoke.

The use of technology to address these issues presents a specific range of questions that practitioners must answer on a case-by-case basis:

(i)   Are parties who use these tools obligated to disclose such use?

(ii) Where disclosure is required, how much information is appropriate?

(iii) If disclosure is not required in a certain instance, might it still be *desirable* in order to best advance the client's interests?

Are parties who use these tools obligated to disclose such use?

If the failure to disclosure the use of such tools could, under the circumstances, be considered unfair to the opposing party, or create a misrepresentation about the discovery process, then the tools must be disclosed under ABA Model Rules of Conduct 3.4(a), in order to ensure the requisite fairness to the opposing party.[3]

---

[1] ABA Model Rules of Conduct, 1.5(a).

[2] Case for Cooperation, The Sedona Conference Journal, Vol. 10, p. 340 (2009) (emphasis added).

[3] As elaborated in Comment 1:

> The procedure of the adversary system contemplates that the evidence in a case is to be marshalled competitively by the contending parties. Fair competition in the adversary system is secured by prohibitions against destruction or concealment of evidence, improperly influencing witnesses, obstructive tactics in discovery procedure, and the like.

Similarly, in the context of a dispute about the discovery process submitted to the court for resolution, failure to inform the court that analytical tools were used could potentially violate ABA Model Rule 3.3, which mandates candor towards the court.

Where disclosure is required, how much information is appropriate?

As much as possible.

This is a glib, but not inaccurate. There is an increasing body of case law that expresses an expectation from the bench that parties and their counsel will, quite simply, be as transparent as necessary to effect the efficient and complete disposition of the discovery process.

In *DeGeer v. Gillis*, the court concluded that, "[s]electing search terms and data custodians should be a matter of cooperation and transparency among parties and non-parties."[4]

As directly stated to the parties in *Tadayon v. Greyhound Lines, Inc.*, "without surrendering any of their rights, [the parties] must make genuine efforts to engage in the cooperative discovery regimen contemplated by the Sedona Conference Cooperation Proclamation."[5,6]

The court in *William A. Gross Constr. Assocs., Inc. v. Am. Mfrs. Mut. Ins. Co.*, found that parties must be transparent about the technological methods used in discovery in order to ensure the defensibility of their processes, and minimize the likelihood of costly discovery related disputes down the road.[7]

The judge specifically concluded that:

> Electronic discovery requires cooperation between opposing counsel and transparency in all aspects of preservation and production of ESI. Moreover, where counsel are using keyword searches for retrieval of ESI, they at a minimum must carefully craft the appropriate keywords, with input from the ESI's custodians as to the words and abbreviations they use, and the proposed methodology must be quality control tested to assure accuracy in retrieval and elimination of "false positives." It is time that the Bar—even those lawyers who did not come of age in the computer era—understand this.[8]

If disclosure is not required in a certain instance, might it still be desirable in order to best advance the client's interests?

There are, of course, legitimate imperatives around the protection of work product, attorney-client privilege, client trade secrets, employee privacy, etc., that must be taken into account when determining how to address technological issues, in the first instance, and further, in establishing the extent of appropriate transparency with opposing parties around these issues.

---

[4]755 F. Supp. 2d 909, 929 (N.D. Ill. 2010).
[5]Civil No. 10-1326 (ABJ/JMF) at p. 11 (D.D.C. June 6, 2012).
[6]See The Sedona Conference Cooperation Proclamation, (July 2008).
[7]256 F.R.D. 134 (S.D.N.Y. 2009).
[8]*Id.*, at p. 136.

But, in order to strike this balance, lawyers must conduct an affirmative analysis of whether there is in fact a *purpose* to eschewing transparency, rather than simply an instinct. As the court observed in *DeGeer v. Gillis*, "candid, meaningful discussion of ESI at the outset of the case," could have averted expensive motion practice and additional discovery processing.[9]

13.3 Beyond eDiscovery: Transparency in Transactional Matters

ESI is not only a factor in relation to eDiscovery, nor is litigation the only practice area in which parties may find themselves in an adversarial posture—or, at least, with interests that are not perfectly aligned. These issues are highly relevant to many other practice areas, ranging from securities law to real estate to mergers and acquisitions, where associated questions of transparency are raised. It is worth noting that the professional conduct rules that govern disclosures in the context of discovery are not always applicable in transactional matters. Nonetheless, the self same ethical rules apply vis-à-vis the efficient disposition of client resources regardless of the nature of an attorney's work, and there are often contractual stipulations regarding candor in transactional matters.

It is, therefore, somewhat misleading that much of the literature in this area focuses on the impact of exploding data sets in the context of litigation. For transactional attorneys, these issues may present themselves in terms of information governance structures or client privacy considerations, but they are no less significant. Whether engaging in pre-merger due diligence, or conducting a voluntary self-disclosure under Office of Foreign Assets Control ("OFAC") regulations, attorneys must consider how best to achieve sufficient transparency to maintain a spirit of collaboration, whether with the acquiring party's counsel, or the regulators at OFAC.

---

[9]755 F. Supp. 2d 909 at p. 930.

Chapter 14
Effective Communication

As lawyers are called upon to manage and supervise increasingly complex technological projects, effective communication is critical. Effective communication and communication protocols may take a series of forms, and levels of complexity, as the specific project and overall organization or business dictate. There are three key components to any good plan, specifically: (i) establishing roles and responsibilities; (ii) sharing knowledge among team members and stakeholders; (iii) efficiently responding to emergent circumstances.

14.1 Establishing Roles and Responsibilities

The first rule in project communication is figuring out who should even receive communications and what form that communication should take. One commonly used tool in tackling this threshold issue is known as a RACI matrix.

RACI is an acronym for "Responsible", "Accountable", "Consult", and "Inform".

Responsible—Responsible parties are those who have a direct role in executing the project, or a particular aspect of the project. Particularly when working on larger teams, there can be multiple people with responsibility for a given task.

Accountable—The accountable party at whose desk the proverbial buck stops. This person has authority to make the highest level decisions about the project's direction, budget, and resourcing. There can only be *one* accountable party. If, as you embark on preparing a RACI matrix, you believe there are *multiple* accountable parties, that is a red flag for the necessity to achieve greater clarity in your project's design (and, perhaps, have some frank conversations with colleagues).

Consult—Those consulted on a project are people whose expertise, whether institutional or substantive, will be needed to successfully complete the engagement. In legal matters, these might very well include technology specialists.

© Springer International Publishing AG 2017
K. Williams et al., *The Legal Technology Guidebook*,
DOI 10.1007/978-3-319-54523-3_14

Inform—People who need to be informed might be members of other departments whose work will be impacted by the project. For example, the implementation of a new timekeeping system will impact all of the billable professionals at a law firm, so it would be appropriate to inform that community of key milestones in the project, and offer demos/trainings, etc., to ensure there is full buy-in and adoption of the new system when it is finally rolled out.

RACI Matrix

	In-house counsel	Outside counsel—eDiscovery director	Outside counsel—senior attorney	Outside counsel—junior attorney	Review tool vendor	Review vendor
Custodian interviews	A	I	R	R	I	–
Data collection	C	A	I	I	R	I
Data processing	I	A	I	I	R	I
Training	R	–	R	R	–	R
Review	C	I	R	R	C	R
Quality control	R	I	C	C	I	R
Production	C	R	C	C	A	R
privilege log	R	I	R	R	I	R

The above sample RACI matrix is modeled on a eDiscovery project. The letter in each cell of the grid indicates what role that team member has in the particular aspect of the project. Note: every team member does not need to have a RACI designation for every task. And, as mentioned, there are multiple "R"s for some tasks, but only one "A."

14.2 Sharing Knowledge Among Team Members and Stakeholders

When working with a team, it's especially important to maintain a common repository of the relevant materials as well as guidelines regarding approved or acceptable modifications, i.e., change logs to document how and why certain protocols changed—whether the impetus for said change is external (e.g., new regulations) or internal (e.g., purchase of a new contract management system). The goal is to provide tools that help ensure consistent work product across the team, and defensibility from both a legal and business perspective in establishing rationales for revised processes.

And, as a many complex legal matters have multiple workflows, it may be appropriate to maintain multiple change logs. It might, for example, be necessary for privilege reasons to establish a "lawyers eyes only" change log reflecting evolving legal strategy, different from a change log shared with vendors and other third parties, representing technology driven process changes.

If a change of protocol occurs, the change log should document: (i) the date of the change, (ii) the reason for the change, (iii) the person who approved the change, and iv) the means by which the change was communicated to team members.

An email chain, we must emphasize, is not sufficient to meet these criteria. This, of course, doesn't mean that email shouldn't be an integral communication tool, just not your primary mode of documenting changes. Protocol or guidance changes should be easy to access by any team member at anytime, and not just those currently on the project, but also readily available to people who come on board at later stages, as well as clients and colleagues who may want to leverage any lessons learned in a given matter against future similar engagements.

SharePoint, or other commercially available tools, can be very effective for maintaining change logs. If you are with a larger firm or legal department, it is very likely you already have a technology platform in place that can be applied to this purpose.

14.3 Efficiently Responding to Emergent Circumstances

Many of the project management tools contained in this *Legal Technology Guidebook* are geared towards the avoidance of crises. Preparing an **escalation protocol** is a recognition that not all crises may be completely averted, but to the extent they are unavoidable, they may be responded to so that they are minimized, rather than magnified or exacerbated.

Have escalation protocols in place for when team members might need to seek more expertise or approval from senior lawyers or businesspeople. Consider who would need to be informed and/or provide approvals for changes to a project's budget, technology solution, or substantive legal approach (Hint: see your RACI).

An escalation protocol must anticipate the possibility that any of these circumstances might require fairly urgent attention and response. To that end, make it clear, for example, that a specific legal support manager may approve additional vendor expenditures of up to $1500, and everything more than that amount must be approved by a *named* director(s). Make it clear, for example, that contract negotiators may grant payment terms of up to 45 days, and anything more than that must be approved by a *named* senior counsel(s).

Chapter 15
Achieving Consistency

We've all heard some version of the old adage that if you present 10 lawyers with the same set of facts you will get 20 legal opinions—at least. There is, without question, an art to the work lawyers perform, but there must also be a science. This is particularly true as teams of lawyers and staff grow, and technology is introduced to the service delivery model. So, as law firms and corporate law departments struggle to maintain their footing on changing technological and economic terrain, it is imperative that practitioners develop strategies to manage the complexity that these factors introduce into their work. One consequence of complexity is, often, inconsistency. In certain contexts, inconsistency may be source of creativity and innovation, but when randomly introduced into legal services it will most likely result in spiraling cost structures, challenges to the defensibility of the lawyers' work product itself, and even imperil outcomes. Developing strategies to impart consistency into your work is, therefore, critically important.

Technology is often leveraged to automate certain processes, and one might think that where there is automation, consistency is sure to follow. This is not, alas, always the case. It is not, one could argue, even *mostly* the case. Where automation interacts with human users there is fertile ground for inconsistency. More perilous because that ground tends to be somewhat obscured from view. Assuming the technology itself functions as intended, inconsistency may be introduced either via the humans who input the system data on the "front end," or via the humans who analyze and use the outputs of the automated system on the "back end."

Initially, practitioners may be reticent to apply project management to work that they don't view as being susceptible to rigorous process beyond the traditional productivity and "good citizenship" oversight with which we are all too familiar. The first obstacle may just be to overcome the suspicion that the same type of approach that helps manufacturers build a better refrigerator cannot possibly help a team of attorneys generate a better work product. The reality, however, is that the

© Springer International Publishing AG 2017
K. Williams et al., *The Legal Technology Guidebook*,
DOI 10.1007/978-3-319-54523-3_15

case law increasingly reveals an expectation of process rigor that includes an emphasis on demonstrating consistency.[1]

15.1   DMAIC

With deepest apologies, it is time to introduce an acronym: DMAIC.

There is not one magical process that fits all legal projects. There is, however, one approach that can be applied across virtually any project that includes types of work that will be repeated over time, whether by one person, or many. This is where the aforementioned acronym comes in. Six Sigma practitioners are inculcated throughout their training in the DMAIC cycle, it stands for Define, Measure, Analyze, Improve, and Control:

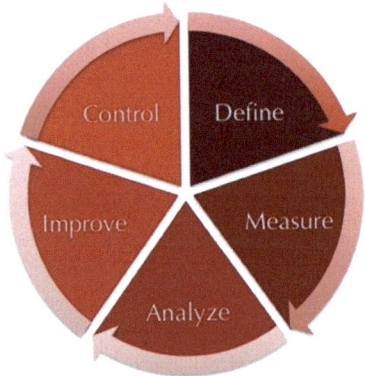

This mantra is meant to evoke a fundamentally iterative approach, which is essential to achieving consistency in any undertaking—while, of course, also helping to assure that the overall quality of the delivery is maximized.

Define

What is called the "definition" phase of any project is often the most crucial, and simultaneously the most often overlooked. Lawyers, in particular, are prone to believe that they already understand the scope of the problem, what they need is the key to solving it … yesterday.

This approach is only compounded in processes like due diligence and discovery related legal project, where the prevailing attitude is often, "I'll know an issue when I see it." Even if this were true—and, often, it is not—there is, as yet, no way to psychically transmit that level of heightened intuition to the teams of attorneys tasked with reviewing the actual documents.

---

[1]See This Chapter—Effective Supervision.

All of this means that it really is necessary to take the time required to set forth the parameters and expectations for a particular project. This will usually entail a series of steps, including:

(i) determining the scope, in terms of both time and resources;

(ii) establishing the project parameters, e.g., will legal staff be tasked with simply identifying documents/issues? Will they be making changes? Contacting counterparties? How will they be expected to interact with the relevant technological tools involved?

(iii) drafting effective training materials that reflect the substantive issues, and that clearly and specifically integrate the protocols and platform functionality set forth above in (ii);

(iv) preparing post training assessment tools;[2]

(v) establishing protocols for ongoing oversight and quality control sampling and;

(vi) implementing documentation and communication plans to track project developments and outcomes.

The idea of spending the time to define these processes may seem anathema. But, all of this can often be accomplished in 24–48 h, and is absolutely essential to achieving an accurate, efficient, and defensible work product.

Measure

You have defined your process(es) and now it is time to measure the project performance against the established guidelines. As the saying goes, "You can't manage what you don't measure." There are few areas where this expression rings so true as it does in legal services. By now, it is routine to have rigorous quality control protocols in place in at least one type of legal delivery: eDiscovery. Sadly, the one exception even in this realm, is often the human, eyes-on review. That is the land where statistical methods and auditable processes go to die. They are sidelined for one-off trainings, ad hoc sampling, and anecdotal analysis. The entire process can break down at this point, rendering moot all of the hard work and planning devoted to effective collection, culling and processing.

Human beings being what we are, there will always be some mistakes. It is, therefore, critical to implement a quality control (QC) process for the project itself that allows supervising attorneys and project managers the ability to identify substantive issues on an objective, timely, and continuous basis. Ideally, you will have prepared a functional QC methodology during the definition phase, *in collaboration with your technology provider/expert*, and any other relevant stakeholders. A successful measurement process will allow the supervising attorney or project

---

[2]That is to ensure that all team members fully understand the training they received and are able to apply their skills effectively.

manager to identify errors on both a team wide and individual basis. Then, using careful analysis and improvement techniques, steps can be taken to drive down the frequency with which these errors occur.

It is usually preferable to establish an initial sampling period of some kind that can be used as a baseline when measuring performance during ongoing QC. This is a ramp-up during which the team can work deliberately to assess if the initial training was sufficient, and address any issues that might not have been apparent in the project planning. During this period the process will, naturally, be less productive, but the gains in efficiency will be recognized in diminished need for rework and rebilling later. It is time well worth taking.

Finally, it is not at all necessary that the project manager be a lawyer. Whether she is or not, it is, however, imperative that she be mindful of the ethical duties arising from the particular legal work being done.[3]

<u>Analyze</u>

After the project manager has identified the various errors and mistakes that occurred during the initial sampling period, they then need to determine *why* these errors occurred. This is a **root cause analysis**, in the Six Sigma jargon.

All errors are not created equal—nor, it is important to note, do they occur with the same frequency. The nature and prevalence and severity of a given mistake will ultimately determine the appropriate steps that should be taken to minimize the likelihood that it is made in the future. Without that key information, a supervising attorney or project manager is operating largely in the dark.

For example, in an eDiscovery review, the miscategorization of not responsive documents as responsive may be caused by something as simple as an issue with the initial set-up of the project platform leading to documents being mistagged. But, it could also reflect a misunderstanding of certain criteria by a specific reviewer, or group of reviewers. This is one of the reasons that it is important to be able to track errors in the measurement process to particular project team members. Not necessarily to penalize those who made mistakes, but because with this information, the project manager can begin to assess whether mistakes appear to be individualized or systemic, which can be a key indicator of cause in some instances.

Just as team members may misapply guidance, guidance may also be inartfully, or incompletely, articulated. The supervising attorneys and/or project managers could discover they need to revise guidance in response to unexpected "realities on the ground." This is often where the aforementioned management and team building skills emphasized in Chap. 14—Effective Communication play a critical role. An experienced project manager can be invaluable in ensuring that all potential causes for errors are fully vetted and explored without diminishing team morale or productivity.

---

[3]See, Chaps. 2, 3, 4 and 10 on ethical duties.

Improve

After completing an analysis of why certain errors have occurred, the next step is to assess the best means to prevent their recurrence. In the context of a legal project, improvement often comes down to training. Training may address issues ranging from the proper use of the project's technological tools, to a substantive primer on the operations of a given business unit. It may be conducted through oral communication, written memo, slideshow presentation, or some combination thereof. Whatever the means, however, there should be some form of documentation created to confirm that the supplemental training took place, as well as who participated in it. Since we are dealing with human processes, it is always helpful to create a record that can be relied on to refresh memories as projects may span many weeks, months, or years.

Also, as with analysis, it is important to recognize that the pipeline for improvement should run both ways. Improvement may require that new protocols be developed, or that new means be deployed to communicate with the project team.[4]

Control

All of the above steps should be continuously applied beyond the first few days of any legal project, just like you should continue to steer your car beyond the first few yards of your journey, even if the road appears to be perfectly straight. Similarly, even if your project has reached a point of maximum achievable accuracy and productivity—legal project nirvana—the only way to ensure that this status will prevail is by maintaining the quality control routine that got you there.

Control may be the most straightforward concept to grasp, but it is almost always the most difficult to execute. This is the point at which parties tend to become complacent, which results in an insufficient allocation (or *mis*allocation) of resources to this critical task.

In the real world, as any legal project progresses novel issues and new tracks of inquiry are bound to be revealed. Each of which may lead to new criteria to define, measure, improve, and analyze. Moreover, if approached diligently, the DMAIC process itself will likely reveal areas were an effective project manager can continue to drive quality improvements.

Accurate processes allow us to minimize the waste of critical resources, which will save money and offer a competitive advantage to early adopters. This is true regardless of whether the resource in question is sheet metal or a billable associate. These processes can also be invaluable as a means of mitigating these risks and driving out the hidden costs of poor quality.

It is almost impossible to imagine a consumer of legal services who might be indifferent to the quality of the work product provided. By the same token, however, it was until recently very common to have no stated expectations between lawyers and their clients (internal or external) regarding the proper management of

---

[4]See, Chap. 14—Effective Communication.

the legal delivery process. As this landscape evolves, there are opportunities for practitioners willing to take affirmative steps to meet—if not exceed—client expectations. Understanding how to achieve consistency within legal service delivery is one way to stay ahead of the curve.

Exercise: Gauging Review Consistency

If a friend gives us a recipe for a casserole that says bake at 350° for 45 min, we assume that his oven's 350° is the same as our oven's 350°. But, why do we assume that? Calibration. Even if you didn't have a name for it until now, calibration is the mechanism that allows us to be confident that each oven coming down the Acme Kitchen Co., assembly line has temperature settings that are in keeping with one another, and with the absolute "truth" of 350°. So, what does this mean for a legal process?

When dealing with problems that require multiple individuals to conduct parallel work streams it is important to calibrate the individuals' work to common decision making. This is not a function of the relative "quality" or credentials of the legal professionals involved. A group of highly educated attorneys are not more likely to *agree* on any individual decision than any other (see, SCOTUS). Therefore, in processes where a measure of consistency is called for, calibration is key.

The first step is to establish a baseline, e.g., what is our true 350°. In most cases this involves identifying a single subject matter expert who is accountable for the content of the work product. After an initial training period to introduce and explain the subject matter expert's approach to various types of issues, the next step is to initiate a Gauge Analysis (**fig**. 15.1) to identify where team members judgments are not in accordance with the judgment of the designated subject matter expert.

This exercise provides a critical and substantive training mechanism, and should be repeated at regular intervals for the duration of a given project to ensure ongoing calibration. After all if, even the highest quality machines fall out of alignment if they are not properly maintained.

Reviewer*	Tag	Doc 1	Doc 2	Doc 3	Doc 4	Doc 5	Doc 6	Doc 7	Doc 8	Doc 9	Doc 10	Doc 11	Doc 12	Doc 13	Doc 14	Doc 15		
Counsel	Responsive	Yes	No	Yes	No	No	No	Yes	No	Yes	No	Yes	No	No	No	No		
Counsel	Privileged	Yes	No	Yes	No	No	No	Yes	No	No	No	Yes	No	No	No	No		
Counsel	Priv Redaction	No	No	No	No	No	No	No	No	No	No	Yes	No	No	No	No		
Counsel	Med. Redaction	No	No	No	No	No	No	No	No	No	No	No	No	No	No	No		
Counsel	Confidential	No	No	No	No	No	No	No	No	No	No	No	No	No	No	No		
Counsel	Post-2012	Yes	No	Yes	No	No	No	Yes	No	Yes	No	Yes	No	No	No	No		
Greg	Responsive	1	1	0	1	1	1	1	1	1	1	1	1	1	1	0		
Greg	Privileged	1	1	1	1	1	1	1	1	0	1	1	0	1	1	1		
Greg	Priv Redaction	1	1	1	1	1	1	1	1	1	1	1	1	1	1	1		
Greg	Med. Redaction	1	1	1	1	1	1	1	1	1	1	1	1	1	1	1		
Greg	Confidential	1	1	1	1	1	1	1	1	1	1	1	1	1	1	1		
Greg	Post-2012	1	1	0	1	1	1	1	1	1	1	1	1	1	1	0	84	93%
Kelly	Responsive	1	1	0	1	1	1	1	1	1	1	1	1	1	0	1		
Kelly	Privileged	-1	1	1	1	1	1	1	1	1	1	1	0	1	0	1		
Kelly	Priv Redaction	1	1	1	1	1	1	1	1	1	1	1	1	1	1	1		
Kelly	Med. Redaction	1	1	1	1	1	1	1	1	1	1	1	1	1	0	1		
Kelly	Confidential	1	1	1	1	1	1	1	1	1	1	0	1	1	1	1		
Kelly	Post-2012	1	1	0	1	1	1	1	1	1	1	1	1	1	1	1	77	86%
Simone	Responsive	1	1	0	1	0	1	1	1	1	0	1	0	1	0	1		
Simone	Privileged	1	1	1	1	0	1	1	1	1	1	1	0	1	0	1		
Simone	Priv Redaction	1	1	1	1	1	1	1	1	1	1	1	1	1	1	1		
Simone	Med. Redaction	1	1	1	1	1	1	1	1	1	1	1	1	1	1	1		
Simone	Confidential	1	1	1	1	1	1	1	1	1	1	1	1	1	1	1		
Simone	Post-2012	1	1	0	1	0	1	1	1	1	0	1	0	1	1	1	78	87%
Tanya	Responsive	1	1	0	1	0	1	1	1	1	0	1	0	1	0	1		
Tanya	Privileged	1	1	1	1	0	1	1	1	1	1	1	0	1	0	1		
Tanya	Priv Redaction	1	1	1	1	1	1	1	1	1	1	1	1	1	1	1		
Tanya	Med. Redaction	1	1	1	1	1	1	1	1	1	1	1	1	1	1	1		
Tanya	Confidential	1	1	1	1	1	1	1	1	1	1	1	1	1	1	1		
Tanya	Post-2012	1	1	0	1	1	1	1	1	0	1	1	0	1	1	1	80	89%
Patrice	Responsive	1	1	1	1	0	1	1	1	1	1	1	0	1	0	1		
Patrice	Privileged	1	1	1	1	1	1	1	1	1	1	1	0	1	1	1		
Patrice	Priv Redaction	1	1	1	1	1	0	1	1	1	1	1	0	1	1	1		
Patrice	Med. Redaction	1	1	1	1	0	0	1	1	1	1	1	1	1	1	1		
Patrice	Confidential	1	1	1	1	0	0	1	1	1	1	1	1	1	0	1		
Patrice	Post-2012	1	1	1	1	1	1	1	1	0	1	1	1	1	1	1	77	86%

Fig. 15.1 Case Study—Gauge Analysis

Chapter 16
Managing Variable Needs

One of the sustaining challenges in almost any business is managing the variability of the business cycle. For lawyers this may mean managing both the variability of our clients' business cycles, as well as the spiky demands created by complex litigation, regulatory investigations, significant mergers and acquisitions, etc. Increasingly, technology, and technology based services, can help attorneys handle the ebbs and flows of their practices. Tools ranging from online onboarding and training, knowledge management platforms, and document control systems allow lawyers to respond to episodic demands more effectively than ever before. Even remote staffing and space on demand, are services that can allow lawyers to scale quickly and efficiently.

Appendix F, to this *Legal Technology Guidebook*, contains an index of specific technology tools and products, but let's first consider scalability as a process challenge.[1] Understanding processes that can improve your ability to scale will, in turn, help you select and manage any technology you may use—whether the "technology" is a pen and paper, or a bespoke intranet platform.

16.1   What Is Scalability?

There are three key components to scalability, as we will discuss it here:

(1) Capability: the capability of a process is its ability to achieve measurable results that meet established requirements or specifications.[2]

(2) Availability: process availability should ensure that necessary resources (human or technological) can be deployed in a timely fashion.

---

[1]Periodic updates to the index may be found at www.legaltechguidebook.com/productindex.
[2]See, also, our discussion of process capability in Chap. 17—Effective Supervision.

© Springer International Publishing AG 2017
K. Williams et al., *The Legal Technology Guidebook*,
DOI 10.1007/978-3-319-54523-3_16

(3) Repeatability: process repeatability is defined as the potential to perform a specified task multiple times over a given period of time.[3]

We will address each of these scalability factors in turn.

16.2 Capability

Capability is like a pipe through which only a certain amount of water may pass every minute. Overburden the pipe, and it will either burst or back up, resulting in a host of expensive damage and necessary repairs. As we have discussed elsewhere in this *Legal Technology Guidebook*, legal processes, like pipes, have their own capability. We can take steps to increase, or limit, a processes capability. But, in order to do so, we must truly understand our workflow, and be able to measure the key factors that may contribute to its success or failure.[4] And, like we want to ensure our pipes can accommodate the volume required, we also don't want to maintain pipes that far exceed even our peak demand—that would be a waste of resources perhaps better allocated elsewhere.

One way to view capability in the context of legal services, is how you handle specialization within your practice. Often, the more experienced attorneys become, the more highly specialized they are within their field of practice. This is typically a necessary evolution in a lawyer's development that also provides great value to their firms or law departments. But, every aspect of a highly specialized practice is not, itself, highly specialized. An experienced "'40 Act" attorney may consult with clients on fundraising strategies, negotiate investor agreements, and directly interact with key regulators. This work, however, also entails research, drafting memoranda routine agreements of varying complexity, and reviewing documents for regulatory requests. Does a highly experienced '40 Act attorney need to perform all of these tasks? Surely, not. A more experienced attorney should, instead, supervise and train junior team members in this work.

The goal should be to assign the varied aspects of the work to the appropriate skill, and by extension, cost level, essentially, disaggregating the service delivery. This can only be reached via a sustained commitment to training—both on substantive aspects of the work for the junior attorneys, and on the art and science of effective supervision, for the senior staff. This will allow greater fungibility across individual staff members, increasing the capability of the process (as well as the availability of resources). Attaining this resource adaptability requires managerial skills that are, unfortunately, not yet part of the core—or, even ancillary—curricula at most U.S. law schools. So, while it is not required or expected that legal professionals have an expertise in these fields. It is necessary that effective training and

---

[3]We refer to "repeatability," but this may also be termed "reliability" in other process management texts.

[4]See, also, our discussion of DMAIC in Chap. 13—Achieving Consistency.

supervision are deployed to achieve scalable processes, which may require drawing on pools of non-legal professionals with diverse skill sets.[5]

16.3 Availability

Maintaining sophisticated technologies in-house, including, expert IT professionals to actually support that technology, as well as keeping up large teams of well-trained attorneys and legal professionals, are all well and good (well and great, even) when each of those resources are sufficiently engaged in revenue generating activity, or are otherwise occupied in ways that support the bottom line of their firm or law department, and the businesses they service. But, for many, if not most, lawyers this sort of full, and consistent, utilization is a much desired, but unattainable, utopia.

In other words, achieving process availability may seem out of reach. After all, cases settle, deals fall through, unexpected crises emerge. External stakeholders, from counterparties, to courts, to regulators may all exert, often unpredictable, influence on the course a legal matter takes.

Nearly all practitioners must, therefor, negotiate how best to allocate their resources to ensure the quality of their services, while maintaining a profitable business. Of course, most readers of this *Legal Technology Guidebook* do not need to imagine dancing on the head of this particular pin—you do it everyday.

16.4 Repeatability

Repeatability is used to determine continuity of outcomes over time. In other words, is your process designed such that the same inputs will yield the same, or materially similar, results. These criteria, or specifications, are defined by you. So, it's important to consider what your specifications are at the outset of your process. Is a lemon materially similar to an orange? Perhaps, but probably not. What about a lime?

These seem like silly questions, but there is a point. Honest. If you have 5 attorneys working on contract origination project, the resources your devote to training them, and—just as importantly—the resources you assign to the quality control of their work, should be determined by the specifications you've set forth. If the insertion of a certain provision is triggered by the presence of citrus fruit, it really doesn't matter if the team members can accurately distinguish a lemon from a lime. It would be a waste of time and money to train them to make that particular assessment, and then to review their work with that differentiation in mind. Alternately, if only oranges and limes, but not lemons, prompt the aforementioned

---

[5]See, Chap. 15—Effective Supervision.

Citrus Fruit Clause, then this would be an appropriate area for instruction, and, quite possible, for quality control.

These same principles apply to the use of automated tools as well. If you are designing a template, for example, over design may be a cause of both unnecessary initial cost, and also undo complexity, which almost always results in costly errors during implementation.

16.5   The Challenges

As noted, above, even if you don't currently spend much time thinking about the scalability of your legal service delivery in quite these terms, it doesn't mean that you, and your organization, aren't confronting these issues every day.

The most common way for most lawyers we've encountered to address scalability is through over-resourcing. Being a belts and suspenders crowd, it is common for lawyers to rely on over-resourcing to address availability challenges. What do we mean by "over-resourcing," though? Simply put, over-resourcing is using staff or assets to complete a project that exceed the needs and/or value of the work.

Examples of over-resourcing might include purchasing a sophisticated predictive coding tool to review 100,000 documents, or assigning an 6th year associate to draft a standard lease agreement. Often over-resourcing is motivated, paradoxically, by a concern for efficiency. That is, thinking that it will be faster to have that 6th year draft that lease, because she's done countless such agreements. A drive for quality may also underpin legal professionals' tendency to over-resource their work. That is, thinking that the sophisticated predictive coding tool will be more accurate than 10 or 20 junior, or temporary, reviewers doing the same work. But, it is important not to conflate "efficiency" with "speed," or "quality" with "credentials."

We are going to repeat this, because it is a fundamental point to this entire *Legal Technology Guidebook:* it is important not to conflate "efficiency" with "speed," or "quality" with "credentials."

Efficiency might increase the rate at which work can be completed, but speed is not, per se, efficiency. It is easy to understand that work done with mindless speed can be filled with costly errors. But, even in our example of an attorney who is faster because she is highly knowledgeable, having completed the same type of work countless times, that speed is likely achieved at the expense of efficiency. Let's do a quick calculation:

Model No. 1	Billing Rate	Hours	Cost
6th year attorney	$550.00	3	$1650.00
		Total	$1650.00
Model No. 2	Billing Rate	Hours	Cost
6th year attorney	$550.00	0.5	$275.00
2nd/3rd year attorney	$300.00	4.5	$1350.00
		Total	$1625.00

In Model No. 1, above, we see it takes our 6th year 3 h to draft this type of agreement. At a \$550/hr billing rate, the work costs \$1650.00. Now, in Model No. 2, we have a 2nd or 3rd year lawyer who, with some supervision, can do the work in 4.5 h—that is it will take the more junior attorney 50% longer to do the work. The junior attorney's time is billed at \$300/hr. All in all, Model No. 2 costs \$1625.00 to complete. At a \$25 difference, this is only marginally less than the cost of Model No. 1. But, we haven't completed our calculations. Consider the 2.5 h that have just been returned to our senior associates. To keep it simple, what if our 6th year simply dedicated those 2.5 h to the supervision of similar work, as set forth in Model 3, below? That would mean she could supervise five additional lawyers, for a total of six person team, thereby increasing not only the efficiency of the process, but also it's scalability and potential revenue.

Model No. 3	Billing Rate	Hours	Cost (6X)
6th year attorney	\$550.00	0.5	\$1650.00
2nd year attorney	\$300.00	4.5	\$8100.00
		Total	\$9750.00

Of course there are other potential variables, and we can't delve into each of them here. But, another less easily quantifiable, yet inherent, risk of over-resourcing, is employee dissatisfaction. The most valuable resources in virtually any legal department or law firm—even, as it so happens, in virtual law firms—are the people who work there. Over-resourcing stifles career development and disincentivizes initiative. At best, attorneys and legal professionals, who are consistently used for work below their real, or potential, skill level will find they are unable to offer the full value to their talents. At worst, they will elect to offer their talents to another employer.

16.6 Solutions

Right-sourcing is a term used to express the correct alignment of resources to a project or process. Unlike over-resourcing, right-sourcing entails using staff or assets whose skill and value/cost are appropriately matched to the work being done. As noted, training and supervision are, without question, critical aspects of establishing and maintaining scalability within legal services, regardless of the available technology. But, understanding the actual value of any specific project or type of work is also very important. Without having conducted such an evaluation, how can you determine whether you are, in fact, using the right resources for the job.

For our purposes, value is not simply a question of revenues, although, it is certainly important to remember not to spend \$1 to make ¢50. Beyond that, however, value includes other considerations, such as, what are, and will be, the core competencies of your law department, and the market identity of your firm. This understanding is crucial in assessing what sorts of skills and technology your organization should invest in developing.[6]

---

[6]See, Appendix D. Case Study—Contract Management Meets M&A.

Chapter 17
Effective Supervision

No attorney is an island. In delivering legal services to clients, many of us routinely rely on fellow lawyers, experts, and administrative support to some degree or another. In doing so, we often use various technology resources to manage documents and other tasks, as well as to enhance communications, knowledge management and scheduling. Of course, as matters increase in complexity, so do the management skills, and tools, required to supervise them.

In this chapter, we are going to address the ways in which legal project management can, and, one could argue, must, be employed for attorneys to fulfill their ethical obligations, specifically, in regards to the duty to competently, supervise junior attorneys, legal support professionals and third party providers. These skills are necessary, and applicable, whether your practice deploys 22nd century technology, or smoke signals and abacuses.

Let's remember that lawyers have ethical obligations to their clients, the courts, and, indeed, opposing parties, that include:

- a duty of preservation
- a duty of competence
- a duty of confidentiality, and
- a duty of supervision.

While this chapter focuses on the duty of supervision, it's worth pausing to emphasize the point that one would be hard pressed to fulfill any of the other duties without effective supervision.

In any given project, practitioners may be called on to supervise an engagement that entails:

- novel and/or specialized technologies,
- multiple subject matter specialties,

© Springer International Publishing AG 2017
K. Williams et al., *The Legal Technology Guidebook*,
DOI 10.1007/978-3-319-54523-3_17

– matrixed stakeholder relationships,
– diverse geographies, and
– outside vendors.[1]

The graph in Fig. 17.1 shows an exponential growth in the numbers of eDiscovery sanctions in the 20 year period from 1990 to 2009. (Note: that we see an increase in the relative frequency of sanctions, not just absolute numbers.) Did litigators become less ethical during that period? That seems unlikely. Instead, one might argue that they became less competent—specifically, less competent in their ability to effectively supervise eDiscovery projects.

While, this may seem like a provocative statement, diminished competence should in no way be equated with a decline in intellectual acumen. Instead attorneys have persistently failed to respond to the algorithmic proliferation of ESI during this time period. So that an approach to eDiscovery—or, for that matter, discovery before the "e"—that would have been competent in 1990, was less so in 2000, potentially sanctionable in 2009, and even more so, today.

No one would rightly accuse lawyers as a class of being inveterate early adopters and, often, the conservative nature of the legal profession, benefits our clients by minimizing risk. But, that is not always the case, and this is a perfect counter example. As you read this chapter (and, indeed, this entire *Legal Technology Handbook*) think about the traditionally siloed and hierarchical ways in which most lawyers work. In your own practices, how safe is it for junior lawyers, or non-attorney legal support staff to escalate issues without being perceived as stepping-out-of-line or not-staying-in-their-lane. The authors will be among the first to acknowledge that lines and lanes exist for a reason, but without clear and open communications, those demarcations will inevitably become fault lines.

If you are unfamiliar with the with the relevant Model Rules of Conduct, now would be a good time to refresh your knowledge of Rules 3.6, 5.1(a), and 5.3, before discussing how legal project management can help attorneys fulfill these obligations. (See, Chap. 3—Outsourcing).

Briefly, Rules 5.1 and 5.3 outline the duty of supervision as it is perhaps most commonly understood, if not applied. Although, non lawyers do not share the

Fig. 17.1 An exponential growth in the numbers of eDiscovery sanctions

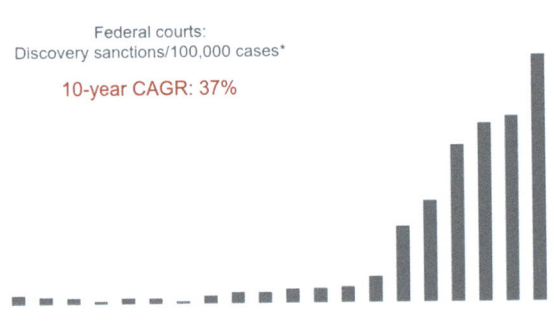

Federal courts:
Discovery sanctions/100,000 cases*

10-year CAGR: 37%

---

[1]Data derived in part from, Dan H. Willoughby, Jr., et al., Sanctions for eDiscovery Violations: By The Numbers, 60 Duke Law Journal 789 (2010); US Court Records.

independent ethical obligations of lawyers, these obligations flow to them via the attorneys who retain or employee them. Further, it is incumbent upon lawyers to supervise non-lawyers in a manner that reasonably ensures these ethical duties are met, i.e., the duties of preservation, competence, and confidentiality.

We've spent some time discussing the evolving standards in Chap. 6, How did We Get Here, and here we will begin to form strategies for responding to this shift.

First, what does "supervision" mean?

Let's start with what it is not: proximity.

Or, certainly not mere physical proximity. Increasingly, effective supervision calls for lawyers to supervise work in which they are not expert—particularly, the use and development of technological tools and products. Some of us are prone to take the concept of oversight a bit too literally. Actually looking over the shoulder of the people working for you is not a recipe for success on any level. Now, this may be particularly painful to come terms with for members of the legal profession who are, notoriously, expert in everything! But, this painful reality calls for a different model from the one in which most of us were trained and still often practice: that is, where the senior attorney delegated work she didn't have time to do competently—or cost effectively—as opposed to work she lacked the specific expertise to do competently. In the case of technological matters, it is almost always true that the lawyers are not subject matter experts.

With that in mind, it is clear that in the evolving technology landscape (if not before) legal project management is, itself, a competency, or field of expertise. Readers may turn to *Apple, Inc. v. Samsung Electronics Co., Ltd*, et al. 11-CV-1846-LHK (N.D.Cal June 20, 2014); *In re Pradaxa Products Liability Litigation*, 2013 WL 6486921 (S.D. Ill) and *J-M Manufacturing Co., Inc. v. McDermott Will & Emery* (Case No. BC462832, CA Sup. Ct., 2011) for examples of how this expertise is often critical to effectuating supervision at ethically and commercially sufficient levels.

In each of these instances, the lawyers did not run afoul of either the courts or their clients because they committed errors—or, not simply because they committed errors. Instead, the failure to adopt robust, documented and effective protocols were the sanctionable/fireable offense in each of these case, and many others.

The following is only a partial litany of the defense counsel's supervisory failures in *Pradaxa*:

Attempting to outsource accountability to:

(a) third-party vendors—as in—production is delayed due to quote vendor issues unquote;

(b) their client's IT department—as in—we told IT to give the vendors full access to the database but for some reason IT provided the vendors with limited access;

(c) their client's employees—as in—our deponent didn't understand that work related day planners should have been produced or the employees didn't understand that work related text messages should have been retained and produced;

(1) asserting that counsels' lack of experience somehow limited their duties of supervision and competence;

(2) claiming that unusual technical issues were at play, i.e., they lacked the necessary expertise to effectively supervise IT support and outside vendors;

(3) claiming that the issues weren't so bad, e.g., yes, we failed to produce this database but it was only 500,000 pages of documents compared to the 3 million we already produced, or, yes that material was accidentally destroyed but the plaintiff doesn't really need it;

(4) counsel tried blaming plaintiffs for submitting too many discovery requests that were broad in scope—but, only after discovery violations were alleged;

(5) counsel claimed—probably, rightly—that they did not know about the problems in their production until they began a comprehensive check or audit of the discovery process in September of 2013.

What's most remarkable about each of these claims, is how unremarkable it is. Collectively, they reflect an entirely commonplace mindset amongst many lawyers, which is—I am only responsible for the oversight of legal issues, with the term "legal issues" narrowly defined to mean things-that-might-appear-on-a-bar-exam. Impatience with this mindset, is precisely the attitude practitioners should expect to see with even greater frequency, not only from courts, but from clients. Project management is quite literally stock and trade for many consumers of legal services, and the are increasingly unwilling to pay for disorganized legal delivery, no matter how brilliant their counsel may be on a given "legal issue".

It is important to put to rest the anxiety-inducing specter of attorneys being forced to attain mastery of ever evolving technology just to fulfill their basic ethical obligations. Everyone loves a good ghost story, we suppose, but that isn't the standard enunciated by any court, nor is it really a commercial imperative set by the market. Instead, the expectation, and requirement, is that lawyers display the minimal levels of management competence that we expect from every other professional field.

Think of the medical profession. Your primary care physician, or general practitioner, may coordinate your care, the delivery of which also may include various specialists, nurses, nutritionists, pharmacists, and physical therapists. It would be unacceptable to the point of absurdity for that general practitioner to wash her hands of referring you to said orthopedist, or reviewing and assessing findings made about your health by said orthopedist, because bones aren't "her thing". Lawyers must, in turn, develop tools to supervise work in which they lack expertise.

We've touched on many of the factors that contribute to making legal project management such a key resource and skill set in a modern practice environment.

But, let's pause to address the elephant in the *Legal Technology Guidebook*: increased financial pressure.

Yes, "Increased Financial Pressure" is a bit of a euphemism when you consider the cost cutting that many practitioners face. Many readers have surely been faced with clients' demands to do more with less. Does maintaining supervisory processes simply add cost that most lawyers can't afford? The short (and accurate) answer is: no. The truth is that effective legal project management can allow practitioners to, first, do better with the marvelous attendant benefit of providing cost savings and predictability.

To use a familiar analogy, the term "fire drill" is often deployed to describe the crisis driven nature legal service delivery. Sadly, that phrase may be even more apt than most people realize. Spontaneous conflagrations of indeterminate cause are rare. Instead, like most fires, crises accrue slowly, and start small. They are only recognized suddenly. Catastrophically. And, like with actual fires, there are also two ways to fight service related crises: prevention and remediation.

Unfortunately, most lawyers still only focus on the latter. But, who wouldn't rather prevent the fire in the first place? After all, prevention is cheaper and less traumatic. The problem is that too many lawyers have positioned themselves as firefighters, or 'remediators.' It's a sexier (and, potentially, more lucrative) gig. Legal firefighters have marathon drafting sessions, make multiple oral arguments, and employ "casts of thousands." Legal fire-preventers have records management, communication protocols, and statistical sampling. They are not the protagonists in best-selling novels.

Legal services consumers, however, are beginning to look at how to keep the fire from getting out of hand in the first place. They're asking about technology and training, planning and protocols, documentation and design. They're expecting answers. In turn, lawyers and legal support professionals need to recognize that legal project managers, while often masters of remediation, have tremendous value to add by preventing the fire drill.

It's not news that lawyers often take a posture of omniscience. This may be a function of the natural disposition of many who enter the profession, or a more carefully cultivated strategic affect—or both. In either event, that pose is less convincing, and less productive (if it ever was) when faced with matters that span multiple geographies, pull in disparate stakeholders, leverage myriad technologies, and trigger critical financial pressures—just to name a very few of the various factors that are often in play.

In this environment, attorneys must rely on the expertise—not just the labor—of many contributors, including, non-lawyers and "junior" team members. And these same attorneys must be able to create a working environment that fosters the enthusiastic, candid, and sustained involvement of those contributors. Trained in the intensely hierarchical institutions of the profession, i.e., firms and courts, this can be very challenging for many lawyers. Not, however, to put too fine a point on it... get over it. Seriously. Those who don't are failing their clients.

Start by remembering that team leadership and team participation are not, and cannot, be mutually exclusive. Next, ask whether it's possible to fulfill your duties of competence and supervision without also being humble enough to encourage and respond to novel, or opposing, views and facts from collaborators at all levels? It's long past time for the lawyers to recognize they have to fulfill a duty of humility if they hope to provide the most effective possible representation for their clients.

17.1 Six Criteria for Effective Supervision

Okay, with that out of the way, let's discuss the criteria for effective supervision.

First: Non-Hierarchical Management—we have to put aside the strictly top down model with which so many of us are accustomed. Look back at our case law examples, in each instance issues arose from the senior or supervisory attorneys lacking insight into the day-to-day administration of the work.[2] None of these were exactly "lightning strikes." They were only unforeseen or sudden crises as a function of poor team management. It is, therefore, critical that the people handling the day-to-day work are able to act as effective issue-spotters. This means the deliverer of bad news must not be met with a storm of arrows. In other words, there should be an atmosphere of trust, rather than recrimination.

Of course, it may be difficult to maintain this standard in what are often stressful circumstances. But, that is precisely the environment in which such standards are necessary, and let's also remember that legal work is ripe for project management, precisely because it is often challenging.

Second: Consistent Communication—one way to minimize the anxiety induced by giving and receiving "bad news" is to have regularly scheduled issue-spotting, or process improvement sessions that include all the relevant stakeholders, such as IT, outside vendors, subject matter experts, a client representative, etc. Doing so helps to establish a culture where getting it right is more valuable than covering your ass-ets, so to speak. Regular debriefs can also help to get ahead of potential crises, and over time you will see a marked decrease in the severity and frequency of "bad news."

Third: Open Communication—it is key that information is not unnecessarily siloed. Why? Well, you wouldn't ask someone to pack your bags for a trip without telling them where you're going would you? So, yes, if there are legal or business reasons to withhold information from team members, then, of course do so. But, the default should be to share information with the people you are relying on to assist you in providing effective counsel to your clients.

Particularly when dealing with technology specialists, a common source of process failure is a lack of clarity regarding project objectives. That is the legal team telling their technology partners the "what," but not the "why." This, effectively,

---

[2]Full text of case law cited throughout this work is available at www.thelegaltechnologyguide-book.com.

limits the ability of experts to act as experts, by finding faster, cheaper, more strategic solutions to problems they are uniquely equipped to tackle.

Fourth: Process, Not Quotas—it is the intrinsic nature of legal work to have to perform against deadlines. Because of this reality, it may be tempting to treat projects as simple challenges of arithmetic: divide the volume of work by the number of working days, and voilà we have our daily/weekly/monthly quota.

But, that approach puts the proverbial cart before the horse. Good project management focuses, first, on what is called process capabilities. In this context, "capabilities" is sort of a term of art, and it's easiest to use an example to help describe the meaning of the term.[3]

Imagine an attorney who has been tasked with supervising a discovery review project. She's identified one million documents that need to be reviewed in fifty working days. That means she has to assemble a first level review team to get through an average of 20,000 documents a day. Based on the attorney's experience with other similar projects, she thinks that each first level reviewer should be able to review 500 documents a day for relevance and privilege. That gets her to a team of 40 first level reviewers. Now, our lawyer has assembled a team of temporary attorneys to do this review. They've been instructed on the issues of the case, trained on the review platform, had an administrative orientation.

So far, so good. But, then our attorney tells her team that, because they are working with a hard and fast deadline, they must review 500 a day, in a 10 h work day. Anyone who fails to meet these targets will be let go.

Our hypothetical supervising attorney has just gone off the rails.

Careful readers may protest here. After all weren't you just exhorted to be as transparent as possible, and here our hypothetical attorney is being transparent.

She's just telling them what needs to get done, and timeframe we have in which to do it. That is an astute observation, but remember that transparency goes both ways. Recall that poor, arrow riddled messenger I mentioned earlier?

What if, for example, there is significant data latency, or complex tagging, or annotation requirements such that a first level reviewer, on average, could only accurately review 400 documents a day? In other words, what if the process capability was less than expected? If that is the case, there are three possible outcomes:

(1) Modify one or all of the process constraints, e.g., identify and resolve latency issues, or streamline the review requirements;

(2) Increase the team size or required work hours, by 20% (assuming, that there would be no diminution in quality as a result—big assumption);

(3) First level reviewers complete 500 documents a day, but do not take the necessary time to accurately assess many of them.

---

[3]See, www.thelegalguidebook.com for process mapping exercises.

Quotas distort incentives such that the overwhelmingly likely outcome is #3. In this example, it is the first level reviewers who will, in short order, have the most insight into the true capabilities of the process, and by instituting a quota, our hypothetical supervisor has diminished the likelihood that she will actually be informed of those capabilities until it is too late—meaning, once a significant quality related issue manifests itself in some costly and/or embarrassing manner.

This is more work for the supervisor than simply managing to a quota entails. Identifying process constraints, setting priorities, and communicating with stakeholders—and sometimes managing up to her own partners or bosses—are all crucial activities. These aren't all skill sets that one can expect to master overnight, or even by reading one excellent chapter on the topic. But, by identifying and understanding what good process management looks like, readers can begin to build those skills and/or leverage existing expertise within or without their organization to fill those gaps.

Now, as a quick note, we should also mention that, quotas can also tend to suppress productivity where actual process capability exceeds the quota levels. In other words, sometimes a quota operates as an artificial ceiling rather than an artificial floor. But, the key word is artificial.

In this example, as in the real world at large, managers often rely on quotas as a sort of proxy for communication. And, therein lies the problem. In order to be truly open—and therefore effective—communication must run both ways.

Five: Establishing Clear Roles and Responsibilities

RACI

As we've seen, part of avoiding process failures is determining who "owns" each given aspect of the project. The RACI Matrix helps to remind us that responsibility and accountability do not always reside in the same person.[4]

To effectively put together something like a RACI, or to do any type of roles and responsibilities planning, the supervisor or project manager will have necessary conversations with the stakeholders that actually serve to help you by giving you a deeper understanding of not just the legal but business drivers motivating certain projects. They will then, typically, be able to anticipate where certain parties may have conflicting expectations, for example, and be able to head problems off at the pass.

Another benefit of this exercise is that, frankly—without getting to 'pop' psychology about it—people just want to feel that their concerns are being heard. You will find much less resistance, particularly when a project requires aspects of how people work will have to be modified to some extent, if they feel that they have had an opportunity to express their concerns at the outset.

Six: Documentation—perhaps a deceptively simple word in this context, since it encompasses such a wide array of work products. In terms of legal and business scrutiny, the focus of documentation should be on transparency, defensibility and

---

[4]See, definition and matrix in Chap. 14—Effective Communication.

repeatability. That means any documentation should advance or support one of these three goals, and if it doesn't, rethink it.

If you consider the RACI Matrix just presented as an example, is a document that it primarily operates under the transparency category. But, depending on the nature of the given engagement, it could also support defensibility and repeatability functions.

Now, think of guidelines or exemplars; most attorneys maintain templates of documents they use routinely in the course of their practice, whether form agreements, document requests, engagement letters, etc. As with any other form of documentation, guidelines and exemplars can—and, typically, should—evolve over time.

When working with a team, it's especially important to maintain a common repository of the relevant materials, as well as guidelines regarding approved or acceptable modifications, in addition to escalation protocols for when team members might need more expertise or senior approval. The goal is to provide tools that help ensure consistent work product across the team. Consistency, of course, helps to drive efficiency by minimizing the need for rework, but it can also speak to the defensibility of a process.

And, finally, along those same lines, quality control processes should be documented. Quality management is, frankly, a huge topic, that encompasses training, sampling, coaching and remediation, as well as the guidelines and exemplars just discussed. But, it's important bear in mind that it is *process*, and one hardly worth the paper it's documented on if team members don't comply. The fact that you have invested in this *Legal Technology Guidebook* means that you are probably already interested in these issues, but others at your firm or law department might not be so keen.

Introduce quality management in stages. As you get a few early wins, so to speak, additional steps will be easier to make, If, however, your quality management process can't be documented, it is not a quality management process.

None of this means there isn't room within the framework of effective supervision—a lot of room—for intuition and informality, ad hoc assessments and offhand appraisals. These all have an important role to play in maintaining a productive work environment and office culture, not to mention being part of normal human interactions. But a process must be repeatable, verifiable, transparent, and documented.

As readers begin to develop both personal and institutional supervision capacity, it is okay to start low and build. Consolidate gains, and then take a next step. In short order you will find you have lots of buy in and institutional support for your efforts.

Chapter 18
Using Technology for Improved Billing and Business

18.1 A Brief History of the Billable Hour

It may seem that the billable hour has been the coin of the realm for legal services since time immemorial, but that is hardly the case. In the United States, the billable hour model only became prevalent in the late 1970s, after the Supreme Court ruled, in *Goldfarb v. Virginia State Bar* that minimum fee schedules set by State and local bar associations violated antitrust laws.[1]

The ruling in *Goldfarb* came at the fulcrum of a period where arrival of the 1938 Federal Rules of Civil Procedure and New Deal regulatory regimes were introducing overwhelmingly more complexity and unpredictability into legal practice, which, in turn, made fixed-fee arrangements less feasible.[2] The convergence of all of these factors lead to the emergence of a common wisdom that hourly billing was not just the preferred model for legal services, but the necessary model.

This "common wisdom" reached its apogee at the beginning of this century, when the ABA proposed a target of 2100 billable/2300 recorded non-billable hours (e.g., pro bono, practice management, etc.) for law firm associates.[3] Both clients and practitioners (excepting, perhaps, law firm equity partners) have been in varying stages of revolt ever since.

---

[1] *Goldfarb v. Virginia State Bar*, 421 U.S. 773 (1975).
[2] Stuart, L. Pardau, "Bill, Baby, Bill: How The Billable Hour Emerged As The Primary Method Of Attorney Fee Generation and Why Early Reports of Its Demise May Be Greatly Exaggerated," 4 Idaho L. Rev 50 (2013).
[3] ABA Commission on Billable Hours Report, 2001–2002.

© Springer International Publishing AG 2017
K. Williams et al., *The Legal Technology Guidebook*,
DOI 10.1007/978-3-319-54523-3_18

18.2 Using Technology to Develop Alternative Fee Arrangements

Through the use of technology, lawyers are now positioned to master the complexity that made alternative fee arrangements seem impractical in the last century. With benefits from doing so redounding to clients, firms, and firm employees.

eBilling Tools

Leading ebilling systems are designed to be more than bookkeeping and invoicing tools.[4] They provide a means of tracking costs by a series of cross referenced metrics, including, matter/deal type, task, timekeeper, counterparty, and so on. Such products not only introduce a degree of cost control and spend management, by offering a granular understanding what tasks are most efficiently handled by which resources, but they also provide an aggregated basis for adopting fixed fee and other alternative fee arrangements (AFAs).

Project Management Tools

Whether used alongside an eBilling system, or on its own, Project Management technology is another way for lawyers to manage and, more importantly, understand, the costs incurred in delivering legal services to their clients. Best in class Project Management tools are designed to track matter progress against user defined milestones. They will also, typically, facilitate collaboration across team members through enhanced communication functions, including, for example, system generated alerts when given tasks have been completed.

If thoughtfully implemented, Project Management tools provide practitioners with critical insight into process capabilities.[5] Users can see which tasks create bottlenecks, and track this data across projects. This information can then be used to revise workflows or redeploy resources more efficiently.

For attorneys willing to take up the challenge these are among the tools available to help them turn the practice, in whole, or in part, away from hourly billing towards AFA models that, if managed correctly, should: (i) provide controlled and predictable costs for clients, (ii) preserve revenues and profitability for outside counsel, and (iii) increase associate satisfaction and retention.

---

[4]See, Chap. 7—Technology Then and Now, for lists of available of e-Billing and project management technology products.
[5]Process capability is discussed in more detail in Chap. 16—Managing Variable Needs.

18.3 Buying and Selling Legal Technology

As we have discussed throughout this guidebook, lawyers are both consumers of technology themselves, and also may be the providers of technology to their clients —whether through developing their own in-house solutions, or working in partnership with outside providers.

As products and providers proliferate across the legal landscape, it is increasingly important for lawyers to be educated consumers of technology. Similarly, as law firms (and, in-house legal departments, vis-à-vis their internal clients) begin to develop and/or bundle technology into their legal service delivery offerings, they must understand the ways in which technology buyers differ from legal services buyers—even within the same company.

The purchase of technology services is often arranged pursuant to highly formalized request for proposals (RFPs). RFPs are usually managed by procurement professionals, who themselves rely on a specialized technology ecosystem to handle all of the product information from, and communications with, would be providers. RFPs are used to ensure, as much as possible, that buyers are able to make "apples to apples" comparisons between various providers, and their complex offerings. They also allow buyers to focus in on the particular functionalities that are most important for that particular law firm or legal department.[6]

Law firms without deep experience either buying complex technology services or, alternately, competing to provide services heavily leveraging complex technologies should consider working with a consultant to help them navigate the procurement process, as it can be quite different from the traditional sale and purchase of legal services. This is particularly true in terms of pricing where hourly billing will often be either not applicable, or not competitive.

18.4 Understanding Business Impact of Competence Failures in Technology

A cornerstone truth of every quality management system is that the customer defines quality. Of course, clever advertising can create an affinity for this brand of coffee, or that breakfast cereal. But, by the same token, if the customer is really craving corn flakes, delivering the world's most perfectly executed eggs benedict won't put a smile on her face.

This is a simple reality legal practitioners sometimes have trouble fully accepting. We lose sight of the fact that if a service either takes longer than the client wanted, or costs more than they budgeted (yes, budgeted!) the service was not a high quality service, and it doesn't matter if every case citation was Shepardized,

---

[6]See, Appendix E. Sample RFP Questions, for examples of general RFP questions that can be used in the procurement of legal technology.

or if all of the legal assistants went to Ivy League schools. We get fixated on perfecting that hollandaise sauce—all the while our client is drumming her fingers impatiently, and calculating how much more this will cost than the corn flakes she wanted in the first place.

Again, *the customer defines* quality. So, assuming the client actually pays for your work, what are the **Costs of Poor Quality** (COPQ)?

Here are a few: Maybe they stay with you because you have institutional knowledge of the business, but wouldn't refer you to peers, or use you on other types of work. Maybe they pay their bill, but only after it's been renegotiated, and always more than 60 days out. Maybe you find yourself allocating a lot of non-billable time to remediate client issues. Maybe you're spending money on a technology tool, or other infrastructure, that doesn't meaningfully support what your clients actually need.

Appendix A
Case Study: Network Security

You are a criminal defense attorney working on a high profile case for your client, Bobby Mans. Bobby has been charged with multiple counts of conspiracy to defraud the government. The Prosecutor's office, however, has offered Bobby a deal for a lower sentence if he provides evidence against the ringleader, Leo Star. For his cooperation, Bobby was placed on witness protection and his identity concealed from the public. You have just received files relevant to the case. It is a beautiful day in spring, so you decide to go to the local café. The café offers free wireless connection to the public within the immediate area. You connect to the café's network and begin to work.

What's the problem?

Upon first review, this appears to be a normal situation. Sun is shining, spring is in the air, and you are enjoying the change in seasons while working. So, why should you, the attorney for Bobby Mans be concerned? Perhaps knowingly or unknowingly, you have connected to an open network (the public wireless connection to the Internet offered by the café). The benefit to you is that you can connect to this network without authentication, meaning there is no security measure prohibiting you from connecting to this network. The con is that so can everyone else in the local vicinity.

Think about your home network. When you setup your wireless home network, did you require a password to authenticate access to this network? If so, why do you think you did that? Yes, perhaps it was to prevent your neighbors from using your Internet and slowing down your connection speed. But, perhaps, it was to prevent hackers from gaining access to your computer by connecting to your network. Remember, a network allows communication from one computer to another computer through the router. When connecting to your network, a hacker can intercept the communication being sent from your computer to the router and gain access to all of the information that you are sending over the network.

Now, let us turn back to our case of Bobby Mans. Leo Star, the alleged ringleader, received information about a confidential informant who was due to testify against Leo in his case. Without the informant, the Prosecution would have to drop its case. Leo, of course, hires a private investigator, Jim Knowsi, to uncover the

© Springer International Publishing AG 2017
K. Williams et al., *The Legal Technology Guidebook*,
DOI 10.1007/978-3-319-54523-3

informant's identity. Through methods unbeknownst to you, Jim was able to determine that you were the attorney for the informant and has been following you for several days. When you began work at the café, Jim was only two seats away. In addition to being an investigator, Jim was also an IT-guru and was able to hack into your network connection without much difficulty. By the time you left the café, Jim had intercepted sufficient information about your client, which was delivered to Leo without delay. Now Bobby and the case were both in imminent threat.

Have you breached your duties of confidentiality or competence?

Assignment: a prominent judge has asked her clerk to research similar cases in which the attorney potentially or allegedly broke the attorney-client privilege by negligently exposing client information over a network that was easily susceptible to hacks. The judge would like a memo (1) identifying any common issues; (2) determining how could the attorney have avoided these issues? (3) assessing what duty does the attorney have to avoid these issues; and (4) asking did the attorney fail to meet that duty in this case?

Appendix B
Case Study: Backdoor Access

Imagine this scenario. You are the commercial attorney for your client, Company A, an international manufacturer with over $40 billion in revenue. Company A is in the market for an innovative software solution that will reduce production costs. Across the table from you is opposing counsel and the Chief Information Security Officer (CISO) for Company B, the supplier of software services. Company B developed a customizable application that it can tailor to the specifications required by Company A in order to streamline the ordering and processing of customer orders.

After hours of negotiation, you finally begin the discussion on warranties. Opposing counsel and the CISO refuse to provide a warranty for backdoor access.

Assignment: Negotiations have been delayed while the parties contemplate this impasse. In the meantime, your client would like you to draft a memo, addressing the following: What implications will inclusion or exclusion of this warranty have on Company A? What are the legal risks associated with this warranty?

© Springer International Publishing AG 2017
K. Williams et al., *The Legal Technology Guidebook*,
DOI 10.1007/978-3-319-54523-3

Appendix C
Scalability Tools

1. Knowledge Management Tools—Knowledge Management includes tools and techniques organizations use to capture, analyze, reproduce and mine information created, owned, or otherwise available to it. Examples of legal knowledge management tools include:

   a. Mitratech.

2. Contract Management Tools—Contract Management includes tools and techniques used to create, manage and track contracts generated and/or relied on by an organization. Examples of Contract Management tools, Include

   a. Agiloft
   b. Apttus
   c. Ariba
   d. Capterra.

3. Onboarding Tools—Onboarding is the tools and processes organizations use to orient, track, and train new hires. Examples of Onboarding tools include:

   a. Enwisen
   b. iCims
   c. PeopleMatter
   d. Silk Road/Red Carpet.

4. Legal Outsourcing Tools/Services—Legal Outsourcing entails the use of tools and services to use resources not owned or employed by the organization to complete certain legal tasks. Examples of Legal Outsourcing Tools and Services include

   a. Huron
   b. Epiq
   c. FTI
   d. Kroll OnTrack.

© Springer International Publishing AG 2017
K. Williams et al., *The Legal Technology Guidebook*,
DOI 10.1007/978-3-319-54523-3

Appendix D
Case Study—Contract Management Meets M&A (Role Play)

PACMan Inc.

Teacher Notes
Review General Instructions and Confidential Facts: 30 min
Planning Meeting: 60–75 min (distribute notice of additional agreements 15 min into session)
Initial Debrief: 30 min
Budget Meeting: 60 min
Discussion Questions:
How many of your priorities were met?
What were the obstacles?
Where did you compromise?
Has your group talked about budgets/costs?
What roles did you assign in the RACI (attached)?

PACMan Inc.

General Instructions

About the Business:

Pacific-Atlantic Computer Manufacturing, Inc., ("PACMan") has been in business for over 40 years. Although, PACMan was originally in the business of building computers, word processors and calculators for business use, not long after the company's founding it also began designing rudimentary puzzle-style games for inclusion in the basic software packages bundled with their computers. By the early 1980s, the complexity and ingenuity of these games had advanced such that PACMan recognized a market opportunity in developing dedicated gaming hardware that would also allow them to separately market the games to beyond PACMan computer customers.

Today, PACMan is primarily a business that designs, produces and markets, video games and video game consoles worldwide. PACMan has become hugely successful in this market, with a market valuation of $11 billion. In the U.S., sales of its games and consoles, account for approximately 35 and 40% of the market,

© Springer International Publishing AG 2017
K. Williams et al., *The Legal Technology Guidebook*,
DOI 10.1007/978-3-319-54523-3

respectively. The legacy "computing" business (including, service and repairs) accounts for just under 12% of PACMan's current revenues. Moreover, the majority of these sales are now concentrated in just a few LatAm countries: Moreno, Redonda, and Salve (a/k/a, MRS PACMan).

The PACMan board has approved spinning off the slow growth computing business into a spun-off MRS PACMan, Inc., which will offer investors stable revenues and regular dividends. They will retain the video gaming business in PACMan, Inc., which will appeal to "growth-seeking" investors.

About the Project:

As a result of the spin-off, PACMan must assign, or split, approximately 60 K contracts to/with the new MRS PACMan entity. These agreements include everything from software licenses, leases, sales contracts, and financing agreements. Assigned agreements will require replacing PACMan with MRS PACMan as the contracting party. Split agreements will effectively create a "duplicate" agreement, for MRS PACMan, of existing PACMan contracts.

The majority of these 60 K contracts will require that the counterparty—landlords, customers, facilities maintenance providers, IT vendors, law firms, etc.—at least receive notice of the assignment or split. In many instances, the counterparty will have to actually agree to the assignment or split. This is a *huge* task, and one that's critical to ensuring the undisrupted operations of PACMan and MRS PACMan.

One of PACMan's corporate Deputy General Counsels ("DGC") has been placed in charge of administering this key project. A senior contract manager with the MRS PACMan sales team is acting as the internal project manager.

The contract manager is responsible for, among other things, coordinating the identification and collection of all active contracts that either need to be assigned outright to the new MRS PACMan entity, or duplicated for MRS PACMan to enter into as a new agreement.

After a streamlined bidding process. the DGC has selected Matrix, LLC, ("Matrix") to review the agreements. Matrix is a *non-law firm* legal service provider In there winning bid, Matrix stated that they would deploy a team of 12 attorneys (including, two team leaders) to complete the initial review of the agreements for notice and assignment requirements in 30 calendar days. Matrix' winning bid also includes the use, and associated costs, of Eptitude Data, Inc. ("Eptitude") for data hosting and access to their online review management application.

Mario, Mario & Kong, LLP, ("MMK") is the law firm that has been engaged to support the in-house team on the spin-off. The DGC has asked one of their partners to sit in on the call.

About the Meeting:

The purpose of this meeting is to set timelines, complete the RACI, establish project priorities, and budget allocation.

Confidential Instructions for the Deputy General Counsel

You are one of the corporate Deputy General Counsels for PACMan, Inc. You joined PACMan as Senior Counsel seven years ago after spending six years in the M&A practice for an AmLaw 100 law firm. You will continue to work fro the PACMan entity after the spin-off/divesture. This project is killing your soul.

First of all, it's not legal practice. If you wanted to manage budgets, projects teams, and third-party service providers, you would have gone to business school.

Secondly, this is a totally thankless job. Ever since this project was announced, you've gotten 10 calls a day from all the different business units complaining about how disruptive the contract collection has been, and they don't want you contacting any of their key clients or vendors with any notice and assignment requests. *They* want to handle those communications themselves ... What could possibly go wrong? There is no question that the notice and assignment process will open up a lot of renegotiations with these counterparties, and your intuition tells you that the more of these the business people handle, the bigger the renegotiation universe will be.

To make things worse, it's beginning to look like the sales organization for the Moreno, Redonda, and Salve regions were playing it particularly fast and loose—deviating from the boilerplate, and putting their own "sweeteners" into deals without consulting legal.[1]

Your greatest fear is that a few of these renegade provisions might actually constitute FCPA violations. You've retained MMK to review some of those agreements, along with email and other materials, to help determine any possible exposure. You have no intention of disclosing the fact of this internal investigation during the call.

The Contract Manager is supposed to be helping you liaise with the business units, and find out exactly what in force agreements are out there, but it doesn't really feel like you're on the same team. The Contract Manager will be transitioning into the MRS PACMan entity, along with the sales group the CM currently supports, and you wonder if this is why you're not getting straight answers to some basic questions about the volume of contracts and where/how they reside electronically on the system.

The General Counsel and CFO are very committed to pursuing alternative fee arrangements (AFAs) with any and all of PACMan's legal service providers. There has been an initiative in place to push AFAs for almost two years, but in reality there has not been a lot of progress made on this front. It is important to demonstrate to the broader PACMan organization that the General Counsel's office has a strong business orientation. This contract review has been the perfect opportunity to pursue an AFA model.

The entire budget for this project—through completing all necessary assignments—is $3.25 million. You selected Matrix to review the 60 K contracts for notice and assignment obligations for a flat fee of $950 K. You stipulated a 30 day

---

[1]See, *Glossary of Terms.*

timeline, because you want this work of your plate as quickly as possible, but in reality the analysis can be completed within 90-days. Any negotiated assignments/splits have a 12-month window for completion.

The Matric bid includes the contract reviewers and the Eptitude review technology. The downstream work (i.e., issuing notices and procuring assignment approvals) will have to be completed out of the remaining $2.3 million. You can't imagine that very many agreements require more than notice, given that most of PACMan's boilerplate agreements provide for assignment with 30 days notice by either party. You expect to come in below budget on this project.

Glossary of Terms:

"Boilerplate"—standard contract provisions.

"FCPA"—the Foreign Corrupt Practices Act criminalizes payments to "foreign officials" that may exert influence over the official(s); gifts, samples, interests in business enterprises, loans on favorable terms, offers of employment, tuition payments, payments to third parties and the like can all violate the FCPA.

"In force"—in effect; enforceable.

Confidential Instructions for The Matrix Representative

You are the Managing Director of Service Delivery for Matrix, LLC. You have been in seat for under six months, and a big part of why you were hired was to grow Matrix' contract management business from it's currently meager base. The way you see it there are three phases to this engagement:

(I) review and analyze the notice and assignment/change of control provisions of the 60 K electronically stored contracts, and capture that information in the Eptitude Data, Inc., application; (II) issue notices to all of the contract counterparties (some volume less than 60 K) where change of control notice is required; (III) request assignment approval from all of the counterparties where assignment approval is necessary (some volume less than the notice number).

Share the RACI with the other meeting participants at the beginning of the meeting.

Matrix has already won Phase I of this project with a bid of $950 K and a 30-day timeline. This bid includes not just the cost of the attorney reviewers, but also the cost of hosting the data and licensing the Eptitude software application you will be using to conduct the review.

You negotiated an $85 K fee to Eptitude for the above services. That leaves Matrix with $865 K. You believe that Matrix can complete Phases II and III for $2,000,000, at a comfortable margin.

Matrix is, historically, an eDiscovery "managed review" provider. Matrix is a market leader in this space. In recent years, however, offshore review providers, advances in technology assisted review, and tough financial times for lawyers at home, have all come together to make document review—even high quality, well managed review like Matrix offers—increasingly commoditized. Your margins for this work have been cut to the bone, and your overall revenue growth for eDiscovery review is almost flat.

Matrix' leadership has honed in on contract review as a growth line of business. Until recently, most of this work been done in-house or by outside counsel. So, on a pricing front your competing with law firms instead of other low cost eDiscovery review providers—translation: you can charge more! You expect Matrix' profit margin for the review and analysis of these 60 K contracts to be almost twice as much as your typical eDiscovery review margins. You have built in a contingency for up to 5 K additional agreements. Beyond that, you will really start eating into you margins.

If you can do the review and analysis with a smaller team, you will be able to restrict the project to your top people, and gain some efficiencies over time. Going from 30 to 60 days will make this engagement $100 K *more profitable* for Matrix (even after paying Eptitude for an additional 30 days of hosting and review application access).

In your eagerness to get this project, however, you may have *slightly* oversold Matrix' previous experience in contract work. It's not that Matrix hasn't done *any* contract review, of course. It's just that, before the PACMan/MRS PACMan spin-off came along, most of your contract work has been on, what you might call "legacy" projects. Basically, taking a population of executed agreements and uploading selected provisions into the client's contract management application. These projects have not had tight time frames, and haven't involved any "downstream" work issuing notices or procuring assignments.

You think you've established a good rapport with Deputy General Counsel, but it's obvious that something is up, since the DGC has invited Mario, Mario & Kong, LLP, to participate in the call. You know MMK would like nothing more than snatch the work for project Phases II and III from Matrix, and you fully expect the partner to lay the groundwork by taking digs at Matrix at any opportunity.

The Matrix Representative's Priorities for The Meeting

(i) Extend the review phase timeline from 30 to 60 days. You want to deliver the highest quality at the lowest over all costs.

(ii) Get a preliminary commitment from the DGC for the downstream work.

    a. An engagement to issue the notices, that would be a *really* big win.

    b. Getting a commitment to handle the assignment requests is a bigger reach at this stage. So, you'll see how the meeting is going before you raise it.

(iii) If you can't get any kind of commitment re Phases I and II, at least hold MMK at bay.

    a. You have two option here:

        i. Kill them with kindness; emphasize your respect for MMK's expertise. Provide a real world demonstration of Matrix project management skills by playing well in the sandbox. You don't know the exact budget, but there is probably enough work for everyone to get a little something—look for ways to collaborate, rather than compete with MMK.

Nice guys finish last; if the firm partner tries to make a play for Phase II and II work on this call, challenge their ability to deliver under any kind of budget constraints, or even to manage the dozens of attorneys required to do the work. If you show any deference to MMK, it will be perceived as a lack of confidence/expertise in contract management work.

Confidential Instructions for the Contract Manager

You have been with PACMan, Inc., for over 18 years. You are not a lawyer, but for the past 15 years, you have worked as a contract manager, supporting PACMan's computing sales force—including, both hardware sales *and* service contracts. You are now the senior contract manager for MRS PACMan in the U.S. You will, eventually, transition to the MRS PACMan entity once the spin-off is finalized.

You were selected to work on this project, because you are one of only two US-based contract managers for the MRS PACMan business. It was clear early on that very few of the major stakeholders actually know how the sausage gets made on the contracting front. This has given you a real opportunity to be a "trusted adviser" on the team and raise your profile in the company. You have really relished this role, and consequently, may have expressed a greater degree of certainty on a few points than was really warranted. A few things that give you pause about completing this project within the allocated $3.25 million budget are:

(1) You're not totally certain how many in force agreements there are.[2] Before this whole process you had really never given any thought to tracking agreements that were no longer in force—and it doesn't look like many of your counterparts in other business units have either.

(2) Plus, you know that your group has an unofficial policy to treat expired service contracts as being "in force," so long as the client does not issue an affirmative termination. You are not a lawyer don't really know what the status of these agreements is. It's probably time to bring it up …

(3) Also, you have been working with the 60 K number because the Deputy General Counsel has been all over you about getting the count. Lawyer's always seem to think that technology is magic. The DGC keeps talking about "the system," like all of PACMan's agreements are in one big solitary data-base. But, as you've tried to explain a hundred times, across the different business units, there are five different contract management software applications being used—not including random shared drives. Plus, there are still paper contracts floating all over the company! Some of these *may* have been uploaded electronically, but there's no easy way of finding out how many duplicates are stored electronically and on paper, let alone which agreements they are.

---

[2]See, *Glossary of Terms.*

(4) Finally, the 60 K number does not include addenda. Just like the original contracts, some of the addenda are stored electronically, and others are not. You have no idea how many of these there are, but your best guess is that there are another 5–10 K of these.

None of the contracts that were scanned in from paper are searchable.

Basically, you think that there is a lot more analysis than the DGC and Matrix are assuming to even figure out what contracts are in force, and whether they have been modified by addenda. If you could have another 30 days to work with IT and the business units, you could really get a better handle on what's out there.

The Contract Manager's Priorities for the Meeting

(i)   It is critical that you get the duplicate issue out there, and addressed. The business unit owners have all been very clear that it would be hugely embarrassing to have multiple notices and assignment requests issued to key vendors/partners/clients.

(ii) It is also important to highlight the "in force" question. Like with the duplicates, the business owners feel strongly that they, it will damage the brand of PACMan & MRS PACMan to send out a bunch of notices and requests to counterparties that are no longer ... well ... counterparties.

(iii) You won't swear to the 60 K number. You felt pressured into giving that number prematurely, but there are too many unknown variables at this point. You think that 70 K is a much more realistic number.

Glossary of Terms:

"Addenda"— plural of "addendum," a change or explanation (such as a list of goods to be included) in a contract, or some point that has been subject of negotiation after the contract was originally proposed by one party. Although they are not always, addenda should be signed separately and attached to the original agreement so that there will be no confusion as to what is included or intended.

"In force"—in effect; enforceable.

Confidential Instructions for Mario, Mario & Kong, LLP

MMK is an AmLaw 50 firm that has represented PACMan, Inc., for over 15 years. You have been working on their matters since you started at MMK nine years ago. Although, you were elevated to firm partner last year, PACMan's primary relationship partner is a member of the firms Executive Committee, who you want, and need, to continue to impress.

The reality is that from the DGC's perspective you are on this call for informational purposes only. MMK has been retained regarding an internal investigation of possible FCPA violations.[3] MMK attorneys will be reviewing MRS PACMan

---

[3]See, *Glossary of Terms*.

generated sales agreements that deviate from corporate boilerplate. That said, if you can get it, you want a piece of the downstream negotiation business—meaning the renegotiations of any contracts where the counterparty pushes back on approving the assignment/split to MRS PACMan without some kind of deal sweetener. Not only would this be a good stream of revenue, but also the kind of work on which your junior associates can cut their teeth.

You understand that the legal landscape is changing, but there's more art than science in the services MMK lawyers ... most of them anyway. So, as much as you would like some part of this contract analysis business, AFA is a total non-starter. The firm is not going to expose itself to that level of revenue risk. If it comes to that, you can offer a 10% discount on all bill rates.

If 10%, or 6 K, of the total number of contracts being reviewed will fall into this category, MMK junior associates, who are billed at $250/h. can handle these renegotiations for a the total cost of $3,600,000 (factoring in a modest premium for senior attorney supervision). With the 10% discount, the cost drops to $3,240,000. If you have the equivalent of 8 full-time associates on this, you can complete the work for PACMan in 12 months.

As appealing as it is from a revenue perspective, MMK is really not interested in getting involved in issuing the notices, or initial requests for assignment. You see that as administrative work, that you will do only if required to maintain the relationship with PACMan.

The MMK Lawyer's Priorities for The Meeting

(i)   You want MMK to review all, not some, of the non-standard MRS PACMan sales agreements. If MMK is being asked to issue an opinion on PACMan/MRS PACMan's FCPA exposure, this is critical.

(ii) Get the renegotiation work.

    a. If Matrix manages box you out, it is critical that you define some role for MMK in any downstream contract renegotiations. If you can be involved in the notice and assignment request projects, even better.

Glossary of Terms:

"**Boilerplate**"—standard contract provisions.

"**FCPA**"—the Foreign Corrupt Practices Act criminalizes payments to "foreign officials" that may exert influence over the official(s); gifts, samples, interests in business enterprises, loans on favorable terms, offers of employment, tuition payments, payments to third parties and the like can all violate the FCPA.

PACMan RACI Chart

Project phase	Responsible	Accountable	Consult	Inform
Phase I				
Phase II				
Phase II				

R   Responsible: those who do the work to complete the task; must be at least one R for every task.

A   Accountable: the individual ultimately responsible for the completion of the deliverable of the task. There should only be one accountable person, and they must sign off on work completed by the R(s).

C   Consulted: those who have input/opinions on how the task is completed, e.g., subject matter experts.

I   Informed: those who are kept informed on project progress.

Appendix E
Sample RFP Questions

General

—Indicate the type(s) of products or services you offer, the number of years you have provided those services.

—For each type of the products or services you offer, describe the specific services you are able to provide and your experience in providing them, including examples of past matters in which you have provided these services.

—Provide three to five client references (including at least one law firm and one corporation) for your electronic discovery services.

—Describe additional value-add services your company provides that the client should consider in the evaluation process.

—What sets you apart from your competition? Is there some aspect of your product(s) or processes that is unique?

Project Management and Quality Control

—Describe your reporting and quality control/quality assurance procedures.

—Describe your communication and escalation protocols.

—Describe what user generated metrics your product tracks.

—Describe what system generated metrics your product tracks.

—Describe what metrics you share with the client and the procedures for interpreting such measurements.

—Will you provide project managers who are dedicated to the client?

—What are the typical qualifications of your project managers?

Partners and Subcontractors

—Identify any on-shore partners or sub-contractors you use to provide your services, and describe the specific services you use them to provide.

—Identify any offshore partners or sub-contractors you use to provide your services, and describe the specific services you use them to provide.

© Springer International Publishing AG 2017
K. Williams et al., *The Legal Technology Guidebook*,
DOI 10.1007/978-3-319-54523-3

Security and Technology

—Describe your security provisions for your offices and physical facilities.

—Describe your security provisions for your hardware (i.e., computers, networks, databases).

—Describe your security provisions regarding access to data and how you ensure only authorized users have access to data and systems.

—Describe your business continuity plan, including frequency of testing and any alternative site locations at which staff may carry out their duties in the event of a disaster at one or more of your normal business locations or attach a copy of your policy.

—Describe the security background checks you perform to ensure the suitability of all staff (full or part-time employees, hourly workers, independent contractors and/or sub-contractors) engaged in providing the products or services contemplated by this RFP.

—What security certifications, if any, does your company hold?

—If you are aware of any data breach over the past 5 years, or if there has been an allegation of a data breach, either by extrusion or intrusion, connected to your products or services, please attach a document detailing the nature of said breach, or alleged breach.

Appendix F
Legal Technology Product Index[4]

PRODUCTIVITY

Office Suite

Word Processors

- Microsoft Word
- Google Docs
- Corel Wordperfect
- Apple TextEdit

Spreadsheets

- Microsoft Excel
- Google Sheets

Email

- Microsoft Outlook
- Google Apps/Gmail

Presentation

- Prezi
- Keynote
- Google Slides
- PowerPoint

Collaborative

- Sharepoint
- Google Drive

---

[4]See, also, Appendix C: Scalability Tools.

© Springer International Publishing AG 2017
K. Williams et al., *The Legal Technology Guidebook*,
DOI 10.1007/978-3-319-54523-3

Enterprise Content Management Tools

– Sharepoint
– Xerox Docushare
– Redbooth
– OnBase
– Unisys
– Filenet

File Sharing

– Dropbox
– Google Drive
– Apple iCloud
– Box
– Mozy
– Carbonite
– SugarSync

Business Tools

Accounting
Billing
Voice
Cable/Internet
Translation
Mobility

Internet Marketing

Client Contact/Video Conferencing
Social Media

– Facebook
– Twitter
– LinkedIn

Message Boards/Client Generation

– Avvo
– LinkedIn

Client-Attorney Matching

– Legal Hero

ORGANIZATION

Hardware

Hosting

- Rackspace
- AWS
- Microsoft Azure

Databases

- Microsoft Access
- Oracle
- MySql

Backup Tapes/Disks and Managing Data

Information Governance

- Daegis
- Zylab
- OpenText
- Epiq

Knowledge Management

- Central Desktop
- Novo

Practice Management

Comprehensive Tools

- Thomson Reuters/WestLaw Firm Central
- LexisNexis Time Matters
- Abacus Law
- Amicus Attorney
- Clio
- MyCase
- Rocket Matter
- Practice Panther
- Credenza
- LegalTrek

Task Specific Tools

- Client Portals (Virtual Law Offices)

<u>Simple</u>

- Total Attorneys
- Direct Law

<u>Complex</u>

- Basecamp
- Onit
- Zoho

Timekeeping

- Sage Timeslips
- LexisNexis Bill4Time
- CosmoLex
- eBillity
- Freshbooks

Document Automation

- Hot Docs
- Thomson Reuters Contract Express
- ProDoc
- Xpress Docs

Contracts

Contract Management

- Apptus
- Novatus
- Contact Logix
- CobbleStone Systems Contract Insights
- Deem Contract Management

Contract Signature

- DocuSign
- Adobe eSign
- Citrix Right Sign

eDiscovery

Managed Services

- Huron
- Kroll Ontrack

Forensics

- AccessData Forensics Toolkit

Comprehensive Platforms and Predictive Coding

- Discovery 5
- kCura Relativity
- Kroll_Discovery
- FTI Ringtail
- Everlaw
- Logicull
- Nuix
- iConnect
- LexisNexis Concordance
- Epiq

Legal Hold Notice Tools Collection, Data Retrieval/Forensics,

- Kroll Ontrack
- Access Data FTK Toolkit

Document Review Staffing/Legal Process Outsourcing

- CPA Global
- Pangea3 (Thomson Reuters)
- UnitedLex
- Mindcrest
- Quislex
- Clutch Group

LITIGATION

Conflict Checks (Most conflict tools are found within PM software)

- RTG
- Legal Software Systems

Trial Related Tools

Settlement

- Picture it Settled

Trial Prep

- Casemap
- Court Days Pro
- DocketLaw
- iTestimony
- The Deponent

Jury Related Tools

- JuryPad
- JuryTracker
- iJury
- iJuror
- Jury Notepad
- JuryStar
- JuryDuty

Trial Presentation

Traditional Tools

- PowerPoint
- Keynote
- Prezi
- Google Drive Presentation
- Slideshark

Modern Tools

- TrialPad
- TrialDirector
- ExhibitView
- Sanction
- Visionary

Transcript Review Apps

- DepoView
- Mobile Transcript
- Transcript Pad
- Westlaw Case Notebook Portable E-Transcript

Witness Apps

- Tablit

Timeline Apps

- Timeline 3D
- Timeline Xpress
- TimeMap

Checklist Apps

- Courtroom Objections

Rules

- OpenRegs
- All Law
- Aptorney
- Fed Courts
- Rulebook

Calendaring

- Court Days Pro
- Docket Law

DECISION MAKING

Simple Strategies (low tech)

Decision Trees

- Workflowy

Flowcharts and Toolkits

- Practical Law (Thomson Reuters)

Checklists

- LexisNexis Automated Forms (Checklists)

Complex Strategies/Data Analytics Tools

Data Visualization

- Ravel Law

Legal Process Decision Making

- Mitratechs Cert. Legal Analytics Partner Program

Alternative Case Theory Modeling/Predicting

- Lex Machina

NEXT PHASE

Computing Power Leveraging

Artificial Intelligence

Power Sources

- IBM Watson
- Google Depmind

Watson Plug-ins

Digital Legal Research Advisor

- ROSS
- STUT

Machine Learning and Language

Speech Recognition
Natural Language Processing

Big Data/Analytics
Hybrids

Efficiency Leveraging

Blockchain
Bitcoin

SECURITY

Network Security

Secure Websites

- SSL Protocol

Anti-Virus

- AVG
- Norton Antivirus (Symantec)
- Avast

Firewalls/3rd Party Defense/WebSite Security

- Sitelock
- McAfee (Intel)

eMail Spam Filters and Attachment Scanner

- Barracuda Spam Firewall
- My Digital Shield
- The eMail Laundry

Encryption
Virtual Private Networks

- PureVPN

Glossary

Alternative Fee Arrangement or AFA Billing structures other than the billable hour model for lawyers and staff

Artificial Intelligence Computerized systems performing tasks normally requiring human intelligence

Blockchain Uses linked blocks of data to allow multiple parties to securely execute transactions without relying on a central "clearing" authority; originally associated with Bitcoin

Bring Your Own Device or BYOD Where employees use personally acquired devices to perform tasks related to their employers and/or clients

Change Log A running log documenting changes to a process and its related protocols

Cloud A network of servers providing storage and software as services that can be accessed from local devices

Confidence interval The range within which a given value is likely to occur

Confidence level The probability that a given value will fall within a specified range, or confidence interval

Costs of Poor Quality or COPQ Identification and valuation of any costs associated with producing and/or delivering products, or services, below the client's quality threshold

DMAIC Iterative process with stages: define, measure, analyze, improve and control

eDiscovery The identification, collection, review and/or production of electronic data pursuant to investigation or litigation

Escalation Protocol Procedure to identify and resolve questions or issues through the chain of command

© Springer International Publishing AG 2017
K. Williams et al., *The Legal Technology Guidebook*,
DOI 10.1007/978-3-319-54523-3

Gauge/Gage Analysis Also known as Gauge or Gage R&R for repeatability and reproducibility is used to determine the (undesired) variability exists within the measurement mechanisms of a process, or system

Information Security Management Systems or ISMS Policies and systems related to data security and information management

International Electrotechnical Commission or IEC An international, non-governmental organization engaged in setting standards for electronic. products, systems and services

International Organization for Standardization or ISO An international, non-governmental organization engaged in setting standards for commercial and market related products and services

Local Area Network or LAN A network of connected computers within a defined physical proximity

Machine Learning The use of statistics to develop iterative algorithms

Margin of error An amount reflecting the likely error rate (as opposed to normal variation) in a statistical sample

Metadata Data that underlies, or describes, other data

Moore's Law Gordon Moore's observation that the number of transistor's per inch of integrated circuit had doubled every year since the integrated circuit's creation, and his prediction (in 1965) that the trend would continue for the foreseeable future

Privilege Refers to a right, residing with the client, to maintain the confidentiality of communications with, from, and in some circumstances, the work product of the client's lawyers, and those working under their lawyers' direction

Process Capability The normal productivity a process can achieve over a sustained duration, at a given level of quality output

Project Management Body of Knowledge or PMBOK® Guide The Project Management Institute's guide to global standards for project management

Project Management Professional or PMP[5] A designation, established by the Project Management Institute, reflecting project management competence and training, as defined by that organization

RACI Matrix A tool to document and define project team roles as: responsible, accountable, consulted and informed

Richness The prevalence of a defined characteristic within a population

---

[5]PRINCE2 (PRojects IN Controlled Environments), administered by APMG in the United Kingdom, is also a widely respected and recognized project management credential.

Root Cause Analysis Identifying the primary or, root, cause of an error, which may be different than the proximate cause

Sedona Conference A research and education organization focusing on developing understanding and consensus around various legal issues including eDiscovery

Server In technology, a server may be a program or device that provides functionality to other programs or devices, i.e., the client

Six Sigma A series data oriented of tools and techniques designed to increase efficiency by improving quality

Statistical significance In statistics, significance means the likelihood that a given hypothetical is true, or how likely an outcome is due to a defined variable as opposed to chance

Technology Assisted Review The review of electronic data or documents using machine learning tools for predictions and categorizations

The Project Management Institute or PMI A membership organization for project management professionals

Wide Area Network or WAN A computer or telecommunications network that spans two, or more, Local Area Networks

Index

© Springer International Publishing AG 2017
K. Williams et al., *The Legal Technology Guidebook*,
DOI 10.1007/978-3-319-54523-3

Legal Citations[6]

---

[6]All case law and rules citations contained in this work may be found in full at www.thelegaltechnologyguidebook.com.

© Springer International Publishing AG 2017

K. Williams et al., *The Legal Technology Guidebook*,

DOI 10.1007/978-3-319-54523-3

The manufacturer's authorised representative in the EU is Springer
Nature Customer Service Centre GmbH, Europaplatz 3, 69115 Heidelberg,
Germany. If you have any concerns regarding our products, please
contact ProductSafety@springernature.com

Printed and bound by CPI Group (UK) Ltd, Croydon, CR0 4YY
28/04/2026
02098474-0001